Behind the Veil

Exploring Remote Sight & Non-Human Intelligence

Jody Swannell

JS
Jody Swannell
AUTHOR

This is a work of non-fiction and speculative synthesis. While it draws upon historical, scientific, and philosophical sources, it also includes interpretive and theoretical analysis. Some ideas presented reflect alternative or emerging perspectives and are intended to encourage inquiry rather than assert definitive conclusions.

All reasonable efforts have been made to credit sources accurately. Any omissions or errors are unintentional and will be corrected in future editions if brought to the author's attention.

To my family and friends who have open minds.

"Get your facts first, and then you can distort them as much as you please."

—Rudyard Kipling, From Sea to Sea and Other Sketches, Letters of Travel (1899)

Contents

Preface

Behind the Veil is about a misunderstood intellectual space at the junction of archaeology, mythology, consciousness, and the history of scientific ideas. Works of this kind are frequently categorized too quickly, either dismissed as speculative or fully embraced.

I wrote this book as an inquiry into patterns in human narrative, the construction of sacred architecture, anomalous historical accounts, and the evolving language of science itself. I'm examining the persistent queries that have accompanied humans across millennia and remain open due to the inherent limits of our current methodology.

The material I've shared here comes from a mix of well-established research and areas that are still being explored or debated. You'll find references to archaeology and ancient engineering, as well as philosophy of mind, quantum theory, and studies of unusual phenomena. These fields aren't equally well-supported by evidence, but they offer different ways of looking at a reality that often crosses disciplinary boundaries.

Readers should approach this book as a thoughtful exploration of interesting anomalies and interpretive questions. Some ideas will be familiar within their fields; others might seem unusual or speculative. The importance lies in how they help us understand ongoing efforts to grasp reality, not in unanimous agreement.

A key question in Behind the Veil is whether certain patterns in myths, ancient structures, and modern strange reports are more than just cultural coincidences. They might, in some way, uncover how we think about our physical reality that current models don't fully explain.

Such questions aren't new. They've surfaced in different guises across philosophy, theology, early science, and modern theoretical physics. What makes the collection and understanding of data different across fields?

For someone new to empirical disciplines, this might mean being comfortable with ambiguity and avoiding the urge to rush to immediate answers. For those familiar with esoteric or symbolic traditions, it could mean resisting the temptation to overinterpret alignment as definitive proof. In both cases, the message encourages patience, restraint, and openness to understanding the bigger picture.

Ultimately, Behind the Veil focuses on the body of inquiry: how people create meaning from incomplete data, how sources of knowledge develop, and why some questions remain even after solutions are presented.

Whether these are seen as mental habits, cultural traditions, or signs of deeper structural aspects of reality, the phenomenon is still worth studying.

My book should be read as an invitation to examine what continues to elude us.

Prologue

A Family Outing Behind the Gates

It was one of those perfect autumn days in Algonquin Provincial Park when the leaves were burnt red and gold and the air sharp with the scent of pine. I couldn't have been much more than eleven years old, sometime around 1985 or 1986, when any weekend trip with my parents still felt like an expedition into the unknown.

My father, a Warrant Officer in the Canadian military, was behind the wheel of our cream-coloured Oldsmobile sedan. My mother sat in the passenger seat, occasionally glancing back at me with that calm, reassuring smile that meant, *trust your father,* he knows where he's going.

I kept asking the same question.

"Where are we going?"

Dad just grinned. "You'll see."

He'd always been fascinated by space and technology. His military career brought him into contact with those systems, but it was the ingenuity behind them that truly held his attention.

I'd seen that fascination before. Once, while driving through a base, he pointed out a large, abandoned building. "See that building?" he said. "That whole place and everyone who worked in it was replaced by one small computer." There was awe in his voice, as he considered how quickly technology could reshape the world.

We turned off the main park road onto a narrower stretch, *Lake Traverse Road*, and soon approached a gate that looked fairly official. My father slowed and spoke briefly to a gentleman at the gate. I don't remember the words, only that we were waved through.

Years later, I would discover that the *Algonquin Radio Observatory* was not open to the public at that time. It was a restricted site managed by the *National Research Council*, tucked deep within the park to avoid human interference. Through military connections or discreet prior permission, my father had arranged for the three of us to go beyond those gates.

That's where I saw them.

Rows and rows of white parabolic dishes lined up against the trees, smaller ones arranged with disciplined precision, all tilted skyward like a silent army. Beyond them, something much larger rose, a massive steerable dish towering above the forest. The 46-metre dish dwarfed everything nearby, a white arc against the autumn sky.

I pressed my face to the window, stunned.

We parked and stepped out, careful to stay outside the fences and buildings. We didn't go to the control rooms or approach the base of the great dish; we just walked around the grounds, and it felt electric.

"Are they talking to aliens?" I asked, half joking, half hopeful.

My mother snickered. My dad explained, in simple terms, how the smaller dishes worked together to monitor the Sun's radio bursts, and how the giant antenna listened far beyond our solar system.

I would later understand that they monitored distant galaxies, quasars, and faint signals drifting through interstellar space. The words were technical, and much of it went over my head at the time, but the wonder didn't.

We walked slowly, taking in the whole situation. Movies like *Star Wars* and *E.T.* had already sparked my imagination with space adventures, but this was different. This was real steel, math, patience, and human engineering tucked away in the Canadian wilderness.

On the drive home, I kept asking question after question. My father smiled the whole way, clearly pleased that his surprise had the intended effect.

That day, beyond gates most people never crossed, my father gave me something no movie ever could: the feeling that space was not distant or abstract, but close enough for ordinary people to reach.

The passion those dishes sparked has never diminished.

Years later, I dug deeper into the place that changed everything for me. I read how the smaller dishes worked together to monitor the Sun's radio bursts, and how the big one listened to distant galaxies, quasars, even molecules between the stars.

The *Algonquin Radio Observatory* was built in the late 1950s as part of Canada's major push into radio astronomy. Those rows of smaller dishes we saw were there to watch the Sun, tracking its flares and bursts that could knock out radio signals or power lines. That solar array ran for decades but was taken down in the 1990s; the rows are gone now.

The star of the show was always the giant 46-meter dish, finished in 1966. Just one year later, it made history: Canadian scientists linked it with a telescope across the country in British Columbia. By recording the same faint signals at the exact same moment and combining them later, they produced images of distant quasars sharper than anyone had thought possible. It was the world's first successful *Very Long Baseline Interferometry (VLBI)*. VLBI is a radio astronomy technique that links radio telescopes across continents to act as a large virtual dish, and Canada beat everyone else to it.

Today, the big dish is still standing and is now run by a private company called *Thoth Technology*. It no longer hunts for quasars; instead, it helps measure tiny shifts in Earth's crust, keeps an eye on satellites and space junk, and sometimes talks to deep-space probes. The wilderness is quieter without the solar array, but that massive bowl still turns toward the sky, still listening. The day trip my dad shared with me all those years ago still remains in my memory.

Figure: *Algonquin Radio Observatory, Ontario (Public Domain image; Wikimedia Commons)* Author=Bquine|Date=2008-07-03 |Permission= |other_versions= }} {{Algonquin Radio Observatory}} [[Category:Radio Observat https://commons.wikimedia.org/wiki/File:2008_ARO_ANTENNA1.JPG

The Story of Arecibo Observatory

Nestled in a natural sinkhole in Puerto Rico's karst hills, the Arecibo Observatory opened in 1963 as the world's largest single-dish radio telescope—a 305-meter (1,000-foot) giant built into the landscape. Funded initially by the U.S. Department of Defence for ionospheric studies, it quickly became a powerhouse for radio astronomy and planetary radar.

Over 57 years, Arecibo delivered groundbreaking discoveries:

- 1974: Detected the first binary pulsar (with Algonquin in Canada for some work, but Arecibo's role was key in pulsar studies), providing indirect proof of gravitational waves—earning a Nobel Prize in 1993.
- 1974 Arecibo Message: Broadcast a powerful signal toward star cluster M13, a symbolic "hello" from humanity designed by Frank Drake and Carl Sagan.
- 1992: Found the first confirmed exoplanets orbiting a pulsar.
- Mapped Venus's surface, confirmed ice at Mercury's poles, discovered the first repeating fast radio bursts (2016), and tracked near-Earth asteroids for planetary defence.

It starred in films like *GoldenEye* and *Contact,* becoming a cultural icon.

Trouble mounted in the 2010s: funding cuts, damage from Hurricane Maria (2017), and earthquakes. In 2020, auxiliary cables snapped, followed by a main one, prompting engineers to deem it unsafe. Before controlled decommissioning, on December 1, 2020, the 900-ton platform crashed dramatically into the dish, ending operations for good.

Part One

SURFACE STORY

Chapter One
Manufactured Reality

Space appears calm from afar. That feeling of calm stays with people, especially after they realize how delicate life in orbit truly is. Even a few hundred kilometres above Earth, survival relies on systems functioning with precision, balance, and ongoing correction.

I grew up in a small northern town where the nights were very dark, and I often looked up at the sky. When I was young, around eleven or twelve, I used to climb the tree in our front yard and jump onto the roof to lie back and watch for shooting stars. I did it often enough that sand started coming off the shingles where I landed. Eventually, the roof visibly wore out, and my dad cut down the tree. I remember feeling more disappointed about losing access to the view than the tree.

The sky seemed endless and undisturbed. Meanwhile, everything closer to me was changing. I didn't have the words for it back then, but I could feel the way things looked different from what was really going on underneath.

Physics has its own kind of weirdness. The deeper you explore, the less certain everything feels, as if observing changes what's there. As they say, the math doesn't lie, but the world it describes never quite aligns.

Now, the world watches with incredible awe as *Artemis II* sails around the moon.

Space Stations

Low Earth orbit is often imagined as a realm of constant daylight, technological abundance, and human mastery. In reality, astronauts aboard the *International Space Station* (ISS) circle the Earth roughly every ninety minutes, experiencing sixteen sunrises and sunsets each day.

Life in orbit is highly restricted by physics, power, and biology. Electrical capacity is limited; even routine appliances must operate within strict energy budgets. Food variety is narrow, relying on a small rotation of shelf-stable meals, supplemented only occasionally by small-scale cultivation of lettuce or cherry tomatoes.

Human bodies decline quickly in microgravity, requiring about two hours of daily exercise to counter muscle loss, reduced bone density, and fluid shifts toward the brain, sometimes referred to by astronauts as "water brain."

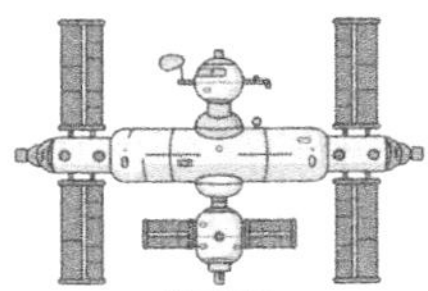

Space Station

Despite common belief, permanent human presence in space is rare. Currently, only two space stations are active: the *International Space Station*, operated through an international partnership led by *NASA* and *Roscosmos*, and China's *Tiangong* station, run by the *China Manned Space Agency*. The *ISS* has hosted continuous human occupation since 2000, but is expected to be retired around 2030–2031.

Tiangong is a unique national project with rotating crews. It aims for scientific growth, exemplified by the *Xuntian* space telescope, a two-metre-class observatory with a multi-gigapixel camera designed to greatly surpass the survey capabilities of the *Hubble Space Telescope*. Other proposed stations, including commercial platforms, remain in conceptual or developmental stages and have not yet reached orbit.

Moon Landing Beliefs

The story of the *Apollo* missions highlights our conflicting impressions. Despite independent verification of physical artifacts on the Moon, some still find the landings hard to believe. The separation between evidence and belief reveals how humans interpret the world. We don't experience reality directly; we experience models of reality shaped by trust, intuition, expectation, and narrative. When an event exceeds what feels plausible, the mind searches for alternatives, even if those alternatives involve more assumptions than the truth.

This reflects a deeper issue in modern physics. At the quantum level, observation is active; our choices, methods, and perspective influence what we discover. The Moon landings have a similar significance in psychology. The data is abundant, yet interpretation varies depending on the observer's initial assumptions. In both cases, reality stays the same, but our understanding of it changes.

Apollo notes that evidence alone doesn't always convince the human mind of the truth, much like how people view historical events differently based on their personal impressions, which influence what they accept, question, or overlook.

Quantum experiments also show that outcomes can depend on how measurements are made.

The Moon landings are an example of how we observe things in quantum physics. They demonstrate that certainty depends on coherence between evidence, expectation, and meaning.

To understand the universe, from astronauts crossing a barren surface to particles popping in and out of existence, we must consider both what is observed and the invisible structures that shape our method of observation. Beyond the hardware, equations, and photographs, the real frontier lies in how reality is understood.

Moon Landing 1969

Few events in modern history are as well-documented as the Apollo Moon landings. More than half a century later, despite overwhelming evidence, the question remains: *Did humans really walk on the Moon?*

The answer is clear: *yes*, based on independent verification, physical artifacts, engineering facts, and international observation. The Moon landings are a triumph of spaceflight.

During the Cold War, the Apollo missions faced worldwide attention. The United States weren't working on it alone. Several countries, including the Soviet Union, monitored the Apollo spacecraft in real time, tracking trajectories, radio signals, and mission timings.

This point is often ignored but is crucial. The USSR had

every geopolitical reason to expose an American deception. Instead, Soviet scientists repeatedly confirmed Apollo's legitimacy. A hoax of this scale would have required a perfect fabrication and the cooperation or silence of America's greatest rival. During a period of intense global tension, this scenario was unlikely by any historical standard.

Apollo astronauts brought back 382 kilograms of lunar rock and soil from six different landing sites. These samples are curiosities stored in NASA vaults and have been studied for decades by laboratories around the world.

The rocks bear isotopic signatures that differ from those of Earth materials, including particles implanted by the solar wind, a consequence of exposure in an airless environment. No known process in the 1960s could have convincingly produced such samples, let alone distributed them globally for independent analysis without detection. The Moon has left its mark on photographs and on parts of itself stored in our laboratories.

During the Apollo missions, astronauts installed laser retroreflectors on the lunar surface. These instruments are still operational today. Observatories worldwide routinely aim laser pulses at the Moon and measure the return time, allowing scientists to determine the Earth~Moon distance with centimetre-level accuracy.

These reflectors are passive, as they do nothing unless struck by a laser. Their ongoing operation is one of the simplest and most elegant ways to confirm human presence on the Moon, since someone had to place them there.

The Apollo program generated extensive documentation: photographs, video footage, telemetry data, mission transcripts, engineering schematics, launch records, and recovery logs. Independent aerospace engineers consistently verified that the Saturn V rocket possessed the necessary capabilities. Its physics are neither mysterious nor outdated; rockets obey the same laws today as they did in 1969.

More revealing than the hardware, however, is the magnitude of the effort. Over 400,000 people contributed to Apollo across government agencies, universities, private contractors, and international partners. Sustaining a deception of this scale for decades would demand near-perfect silence from hundreds of thousands of participants, something human history has never achieved.

Since the Apollo missions, orbiters from various countries, including Japan, India, China, and NASA (USA), have captured images of the lunar landing sites. These photos show descent stages, rover tracks, footpaths, and scientific instruments exactly where mission records indicate they should be.

It's worth noting that these confirmations aren't based solely on American sources. They come from independent space agencies that have no obligation to support U.S. history and, in some cases, might even have reasons to doubt its accuracy.

If the evidence is so persuasive, why does disbelief still persist? The answer lies more in psychology than in science. Many people start with a reasonable assumption: governments sometimes lie. From there, it is easy to believe that any extraordinary claim might also be false. The issue is that Apollo was not a secret program. It was one of the most publicly scrutinized, internationally examined engineering projects ever undertaken.

Common objections, such as the absence of stars in photographs, the "waving" flag, unusual shadows, or the Van Allen radiation belts, stem from misunderstandings of physics and photography. Camera exposures were set for the bright lunar surfaces rather than for distant stars. The flag moved because it was twisted during placement, and without air resistance, its motion carries on longer. Shadows converge due to terrain irregularities and wide-angle lenses. Radiation belts pose real risks, but short transits, combined with shielding, kept exposure within survivable limits.

Modern devices seem more powerful than Apollo-era computers, which might suggest that the missions were impossible. However, this comparison is misleading. The Apollo Guidance Computer wasn't a smartphone; it was a specialized system designed for specific tasks, and it executed them exceptionally well.

Believing in a hidden truth can make people feel stronger. It gives them a sense of identity, a sense of community, and insights others might not have. Because of this, conspiracy theories often satisfy emotional needs, even if they aren't backed by facts. The Moon landings show us something bigger than just astronauts walking on another planet. They reveal how easily stories can hide the evidence, and how amazing truths can seem less real than incredible lies.

Artemis II

The recent *Artemis II* mission complicates the conversation around what we think we know about the Moon. In April 2026, astronauts travelled farther from Earth than any humans had since the Apollo era, completing a 10-day journey around the Moon and safely returning with new data, imagery, and system validations for future missions.

Unlike Apollo, this mission wasn't about landing but about using the technology and human capabilities needed for a sustained lunar presence. For some, *Artemis II* reinforces the official narrative, demonstrating that deep-space travel is not only possible but also advancing. For others, it raises new questions: if we are only now "using" systems for long-term lunar exploration, what exactly was achieved half a century ago?

Whether people see it as a continuation or a contradiction, Artemis II has definitely brought the Moon back into the spotlight as an exciting and ever-changing frontier.

* * *

In a universe full of black holes, quantum entanglement, and unseen forces shaping reality, the Moon reminds us that sometimes the most astonishing thing is that something so *thoroughly* documented can still be questioned.

This really shows how different space is from what we often imagine it to be. Earth's orbit is a very delicate and limited place that needs a lot of work to keep humans alive. If even our most advanced technology struggles to operate a few hundred kilometres above Earth, then we should carefully reconsider the assumptions about where and how more advanced activities might take place.

At first, space, movement, and time might seem like simple parts of reality. However, they could actually be caused by something deeper, an idea that is often misunderstood or ignored. In modern physics, this view is a math-based way to understand black holes, quantum gravity, and the structure of spacetime.

This is a change currently underway in physics: we're moving from seeing the universe as a collection of objects moving through space and time to viewing it as a system of evolving information and relationships, from which space and time themselves spring forth.

This matters because many of the phenomena examined later in this book, such as nonlocal perception, anomalous motion, and the apparent separation of consciousness from physical location, only seem impossible when we assume that space, distance, and time are fundamental and absolute.

In everyday life, particles move along paths, barriers stop movement, and information spreads locally through space. This view works well for most human experiences, but when physics is pushed to extremes, such as near black holes, in the quantum vacuum, or at the universe's beginning, it starts to fail.

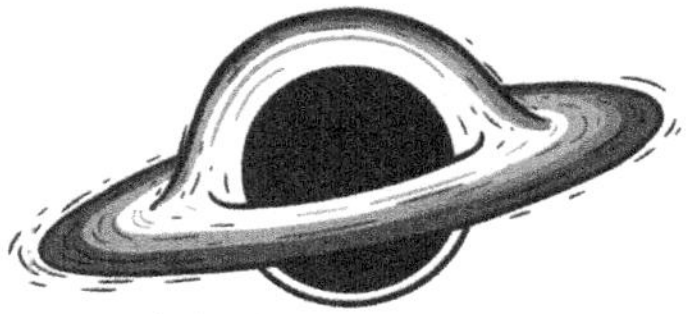

Black Hole with Accretion Disk

Holography

Holography provides a clear example of two ways to view the same system. In some theories, a system with gravity can be described in two ways: as a higher-dimensional universe with space, motion, and forces, or as a lower-dimensional universe without gravity, in which only quantum information follows strict, predictable rules. These two descriptions aren't truly different theories; they are mathematically equivalent and just explained differently.

The idea is that space might emerge from more basic elements. In this view, nothing is actually moving through space, and no distances are travelled. Instead, there are only quantum states, their relations to each other, and the patterns of their change. When these patterns are translated back into the more familiar picture, they appear as movement, separation, and position.

This distinction is important when we see phenomena that

seem to ignore usual boundaries. Imagine a particle on the far side of a barrier. Normally, we think it must have gone through the barrier. But from a holographic point of view, crossing isn't fundamental. The key idea is how information changes over time. The appearance of movement is just a reconstructed effect, not an actual crossing.

This idea appears again in discussions of remote viewing, where information seems to be accessed without a traditional signal between observer and target; in reports of unidentified aerial phenomena, where objects seem to ignore inertia and acceleration limits; and in studies of consciousness, where awareness, memory, and experience don't clearly connect to physical objects. In each case, the physics isn't being violated, but the explanation uses the wrong level of detail.

Over the past twenty years, physicists have gathered increasing evidence that *entanglement, a uniquely quantum form of correlation,* is essential for solving these puzzles. In holographic theories, the shape of space can be deduced from entanglement patterns. Systems with strong entanglement behave as though they are close together; those with weak entanglement act as if they are far apart.

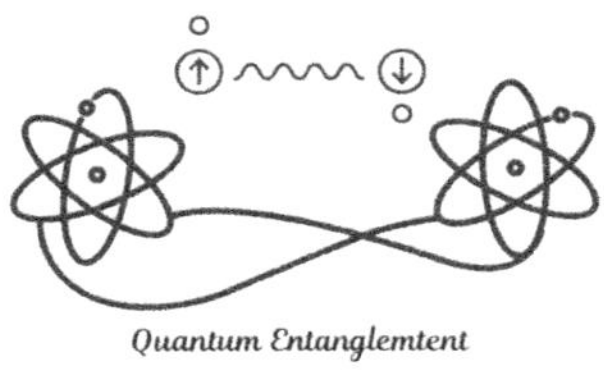

Quantum Entanglemtent

Time follows a similar logic. At the deepest level, the fundamental laws of physics don't include a universal "now." Time arises from how entangled parts relate to one another, changing in response to change.

The flow we perceive is not imposed from the outside on the universe; it originates internally through the growth and reorganization of entanglement. This has direct implications for how memory, observation, and conscious experience are physically realized, topics discussed in later chapters.

None of this suggests that reality is subjective, that observing something creates the universe, or that anything goes. The rules governing these processes are strict and testable. Energy is conserved, causality stays consistent, and information isn't created or destroyed. Emergence occurs when familiar patterns arise from deeper, well-defined rules.

Chapter Two

UAP Activity

Kennedy Space Centre

It was early spring 1989 as the bus rumbled across the flat, sun-baked landscape of the *Kennedy Space Center*, with the horizon stretching endlessly in all directions. Far in the distance, the Vehicle Assembly Building loomed like a giant sentinel, dwarfing everything nearby.

The guide's voice, brimming with excitement, echoed through the bus speakers. "Even from here," he said, "you can see it's still quite a ways off. The flat terrain makes it look even larger. Each star on the American flag painted on the side is about 6 feet wide!" he beamed.

I leaned towards the window, squinting to make out the scale, as the enormous building stood alone like a monument to human ambition. Next, the bus slowed as we passed a marshy stretch of land, and the guide pointed to a high perch in a large tree that stood out against a flat landscape: a massive bald eagle's nest.

My parents and my childhood friend turned to look at it,

one of the great birds standing proudly against the sky, surveying the marsh below as if it were the true ruler of this flat, wild expanse. The contrast between nature and human engineering was almost surreal.

* * *

Later, inside the Visitor Centre, we gathered around a small barricade surrounding the cone-shaped *Apollo Command Module* test article. The guide pointed to its metallic surface and spoke of the astronauts who had sat inside, wearing their white pressure suits.

Although we couldn't get close enough to touch it, the capsule's compact, blunt cone shape made the experience feel intimate and almost tangible. I imagined what it must have been like to sit in that cramped space, with the roar of engines below and the emptiness of space ahead, and felt a mix of awe and reverence.

By Bill White - https://images-assets.nasa.gov/image/KSC-20170912-PH_SWW01_0386/KSC-20170912-PH_SWW01_0386~orig.JPG, *Public Domain*, https://commons.wikimedia.org/w/index.php?curid=113451989

Some nights, I still look up and feel like the sky is empty.

Occasionally, a light moves in a funny way, or a strange shape catches my eye. At first, you just notice it, but then it draws you in, as if the sky's hiding a little secret just for you.

* * *

I often think about Roswell, New Mexico, where a rancher found debris in a field, bizarre enough for the Army to label it a "flying disc." Newspapers covered the story, but the explanation changed quickly. The debris was dismissed as an ordinary weather balloon, and the story moved on, yet the public's shock had already taken root.

The lore surrounding Area 51 grew out of this incident. Fences, hangars, sensors, and the idea that someone, somewhere, is controlling what we're allowed to know. There were a few leaks, partial accounts, and a lot of silence. People filled the void themselves, imagining shapes, lights, and patterns.

I also think about what's going on in Canada. Over decades, thousands of reports document lights and movements of unknown objects in the sky. Civilians observe phenomena that pilots, police, or scientists can't immediately explain. Some reports disappear, some remain in archives, and a few never get resolved. Keeping track of those accounts is what truly matters.

Discover how the world reveals itself to us as we observe, and notice how our perceptions change when small details make us stop and think. The sky keeps changing, ignoring the rules we assign to it.

The more you notice, the more you realize something else: the things we see and can't explain are always filtered through our own perspectives. Observation isn't objective; our eyes and assumptions shape what we perceive.

My thoughts dwell on the strange lights and shapes that defy explanation, making me question how much of our world we truly understand. I also think about how our perceptions influence the stories we tell ourselves about what's possible.

The Roswell Incident

Area 51 is often seen as an isolated, hidden basin of salt flats and hangars, protected by fences, sensors, and armed patrols. However, its true purpose may be influenced by more than geography. From the beginning, its location depended on something less obvious: controlling speculation. To see this, it helps to begin in New Mexico.

In July 1947, a rancher near Roswell discovered debris scattered across a remote field. Foil-like material, lightweight beams, and fragments that were difficult to explain. The nearby Army Air Field issued a brief press release stating it had recovered a "flying disc."

Within hours, the story spread across the country. Newspapers ran bold headlines and radio hosts speculated openly. For a short time, the idea of non-human technology was considered official. *Then the story unravelled.*

The military retracted their statement and replaced it with a far more mundane explanation: a weather balloon. Photographs were staged, and officers posed with the wreckage, which looked like ordinary aluminum foil. The story was considered finished, but the damage had already been done.

Roswell created ambiguity. A mismatch between the initial claim and what was later revealed. Over the following decades, that story was filled with conjecture, testimony, reinterpretation, and myth. Roswell became an example through which broader anxieties about secrecy, technology, and authority were perceived.

Researchers, hobbyists, skeptics, believers, abductees, and debunkers all converged on the same unresolved conclusion.

Conferences were held, archives compiled, and a subculture developed. Folk classifications of flying crafts and non-human beings proliferated. There was a persistent sense that something vital had been withheld, or worse, altered.

By the time Area 51 became public knowledge decades later, the system was already in place. Denial, leaks, ridicule, partial disclosures, and technical explanations convinced some audiences but alienated others. As with Roswell, multiple conflicting versions of the truth existed, none of which were definitively proven false.

Coincidence?

Just a few months after the Roswell Incident, *President Harry S. Truman* signed the *National Security Act of 1947*, a sweeping reorganization of the United States' defence and intelligence apparatus. The act created the *Central Intelligence Agency* (*CIA*), tasked with gathering and analyzing information on foreign threats, and formally established the *United States Air Force* as a separate branch of the military.

This legislation marked a turning point in the U.S. response to conventional and unconventional threats. Amid the Cold War and after unexplained events such as Roswell, these new institutions were established to respond to foreign adversaries and to investigate and contain phenomena that didn't fit neatly within existing military or scientific frameworks. The timing later fuelled theories that Roswell and other unexplained sightings were part of a larger narrative shaping national security priorities.

The *National Security Act of 1947* laid the groundwork for what some historians and researchers now refer to as the CIA's "*legacy programs*." Although the public would not learn of programs such as *Project Grudge* and *Project Blue Book* until later, these early investigations set the stage by establishing protocols, channels, and a mindset for handling unexplained incidents. In this sense, Roswell served as a catalyst that helped shape an entire system of observation, containment, and interpretation.

Holography is a useful metaphor here. You don't see the actual objects, only the information that reaches you. What you believe is happening depends on the pieces you're given and how you assemble them. When people receive different parts of the same story, they naturally imagine very different realities behind it.

Area 51 operates in a similar way. Its secrecy conceals and shapes what people believe is happening. Official silence creates a scarcity of data. Limited leaks and carefully selected details influence how people interpret events.

Media and pop culture both focus on some stories while ignoring others. Based on insufficient evidence, such as radar reports, witness accounts, or redacted documents, people imagine different scenarios: secret planes, foreign drones, military tests, or alien spacecraft.

An interesting question is: how is the knowledge controlled? What determines which interpretations are considered plausible, which are dismissed, and which persist despite decades of doubt?

These questions will come up again in later chapters, especially when remote viewing prompts us to consider whether information is limited to local sources, when UAP reports challenge conventional notions of motion, and when we reconsider consciousness as an active part in shaping reality.

Roswell showed what can happen when you contradict rules about sharing information. Similarly, Area 51 knew that control is about managing how people see things by setting rules.

UAP (Unidentified Aerial Phenomenon)

Long before modern UAP research became a serious topic for science and government, everyday Canadians looked up and reported what they saw. For decades, sightings across Canada have been collected, catalogued, and studied by volunteer researchers and civilian networks eager to understand unusual aerial events from coast to coast.

Since 1989, groups such as Ufology Research of Manitoba have documented tens of thousands of reports (over 24,000, according to recent figures), creating one of the longest-standing national databases of unidentified aerial phenomena outside government archives.

In recent years, this activity has remained steady. In 2024 alone, more than 1,000 reports of unusual aerial sightings were submitted across Canada through official channels and civilian groups, with witnesses describing lights, shapes, and movements that didn't match with known aircraft or natural phenomena.

Previous years also showed similarly high totals: in 2023, researchers documented nearly 600 sightings nationwide, with patterns concentrated in more populous provinces such as Ontario, Quebec, and British Columbia, although sightings were reported in all provinces and territories.

Independent public surveys suggest that between 600 and 1,000 UAP sightings are reported annually across Canada, with roughly one in four Canadians reporting that they have seen something in the sky they couldn't readily identify at some point in their lives.

These numbers show that people frequently observe unusual aerial phenomena and that reporting is inconsistent: only a small portion of those who witness something unusual file a report, and many incidents likely go unreported. Some reports come from civilians, while others are submitted by pilots, law enforcement personnel, and other trained witnesses. Each report adds to the statistics that future researchers try to interpret.

Because there isn't a single national reporting authority and sightings may be collected by multiple organizations, compiling accurate totals is inherently challenging; however, the overall picture is clear: Canada reports *hundreds of UAP sightings* each year, with several recorded daily across the country.

Unidentified Aerial Phenomenon

Placing the Canadian experience in a global context shows how widespread these phenomena are, without assuming any specific explanation. In the United States alone, decades-long civilian databases have recorded tens of thousands of reported sightings, with annual totals in the low tens of thousands.

In Europe, national investigations such as France's *GEIPAN* catalogue thousands of cases, although only a small percentage remain genuinely unexplained after analysis. Similar agencies exist in the United Kingdom, Australia, Brazil, and other countries, each contributing to the understanding that sightings are a nearly universal human experience.

Despite significant fluctuations driven by public awareness efforts, media coverage, and reporting networks, it is clear that many individuals worldwide observe aerial phenomena they cannot readily identify. Only a minor percentage of these reports defy explanation through ordinary factors.

These statistics don't resolve the question of what UAPs *are*, but they do highlight the cultural and informational environment in which the Roswell–Area 51 perspective continues to shape perception. Roswell established a lasting divide between initial claims and official retractions; Canadian and global reporting data indicate that unresolved sightings still fuel curiosity and speculation because official channels often lack the mandate or mechanisms to interpret them definitively.

Depending on the person's mindset, when hundreds of reports, collected over decades, leave even a small percentage of cases unexplained, those unexplained cases become fertile ground for narrative, fear, wonder, or belief.

As with Roswell, the ongoing stream of sightings shows that a significant amount of human observation doesn't fit into known categories. Whether these reports eventually lead to conventional explanations or inspire new ideas in aerospace, optics, atmospheric science, or even in our ways of thinking, the fact remains that ordinary skies produce extraordinary things.

Over time, the way unidentified aerial phenomena are reported has changed as much as the phenomena themselves appear to have. During the early Cold War era, reports were primarily a military matter. Sightings were viewed through air defence, radar anomalies, and possible foreign technology.

Classified Projects

Roswell had barely faded from the headlines when the newly formed CIA began secretly establishing what researchers now call its *legacy programs.* These programs were designed to monitor, catalogue, and analyze unexplained aerial phenomena.

Shortly thereafter, the Air Force launched its own efforts: *Project Grudge* in 1949, which largely dismissed UFO reports as misidentifications or hoaxes. Later, in 1952, *Project Blue Book* was established as a more systematic program to investigate sightings and assess potential national security threats.

Together, these initiatives marked the first organized effort to transform a chaotic and mysterious sky into something the government could monitor, study, and control. What had begun as scattered reports and sensational headlines was now being integrated into a system designed to observe the unknown while keeping the public in the dark.

As official military interest waned publicly in the late twentieth century, responsibility for documentation became a private matter. Civilian groups began to fill the void left by government withdrawal. Volunteer researchers, regional databases, and international networks began, driven by a sense that something important was being overlooked or quickly dismissed.

These civilian efforts notably increased the number of reports and broadened their variety. Reporting standards varied, and witness interviews ranged from detailed to informal. The result was a large archive that, despite its volume, was erratic, generating both interest and skepticism.

Most scientific institutions remained largely absent during this period. The subject carried reputational risks, and without access to classified sensor data, researchers had little incentive to take it seriously. When scientists examined UAP reports, they typically sought to explain them through atmospheric optics, astronomical misidentification, or human perceptual errors.

While the Air Force handled *Project Grudge* and later *Project Blue Book*, other parts of the government and private defence companies were also, invisibly, involved in national security. Official records confirm the government's focus was on observation, analysis, and threat assessment, but rumours persisted of something more: recovered craft, reverse-engineering efforts, and highly classified experiments.

Whistleblowers such as *Bob Lazar* and, decades later, *David Grusch* have claimed that multi-layered programs existed, involving secret locations and contractors. However, because these programs are classified, these accounts remain unverified. The combination of fluctuating agency oversight and classified defence projects continues to linger at the edges of government archives and the public imagination.

In recent years, governments have re-entered the conversation, with a noticeable shift in terminology. "UFOs" became "UAPs," and later "UAPs" expanded to "unidentified anomalous phenomena," subtly broadening the scope beyond aircraft.

Military reporting initiatives now focus on sensor fusion, integrating radar, infrared, satellite, and pilot observations. They openly acknowledge that some incidents remain unexplained due to limited data.

At the same time, scientific interest has been rekindled. Standardized data collection has been proposed, focusing on developing better instruments and clarifying definitions. In this new system, "unidentified" isn't necessarily "alien"; it indicates incomplete information.

Civilian reporting has also improved in response. While

early UFO culture was largely driven by anecdotes and stories, modern civilian databases now focus more on classification, cross-checking, and statistical analysis. Public awareness of drones, satellites, and aerospace testing has grown, helping reduce some misidentifications and drawing more attention to the remaining anomalies.

Over time, different groups have taken on the role of observing, reporting, and understanding strange events. The military looks for threats, scientists are careful with their methods, and civilians share information openly. Each sees the same sky from a different perspective.

This evolution reflects the pattern we saw at Roswell and later at Area 51. The management of UAPs depends on which interpretive systems are activated, which questions are permitted, and which uncertainties are left unresolved. As reporting mechanisms evolve, the shape of belief changes.

This chapter highlights that people around the world continue to report hundreds of unexplained sightings each year, most of which have common explanations, though some remain unresolved, keeping curiosity alive.

Chapter Three

Unfolding Disclosure

There's a moment after you've learned enough to know you can't go back, but not enough for things to feel resolved. I used to believe that disclosure meant simply revealing information as a statement in which someone confidently steps forward and says, '*Here it is.*' It represented a line crossed, with an outright delivery of truth.

However, as I've come to understand how information reaches us, I've realized it doesn't follow a simple sequence. Initially, there's the uncomfortable feeling that what you're observing doesn't align with your usual perspective. Evidence piles up without an understandable explanation.

What's fascinating is how often the issue isn't about the proof. It's about our expectations of what evidence should look like. We all have an instinct to protect our understanding of how the world works. When this view is challenged, we don't automatically dismiss the information; instead, we tend to set it aside or forget it.

Events are gradual processes composed of small truths,

moments when it's acceptable to dive deeper, ask different questions, and acknowledge uncertainty.

This chapter explains why disclosure doesn't happen all at once and why it might not even be possible. It also demonstrates how understanding changes as new information comes to light.

Before delving into institutions, reports, or timelines, it's worthwhile to reflect on how much of our reality relies on the models we accept as normal. How frequently do we confuse familiarity with truth, and how rare is it for us to challenge what is considered reasonable?

Uncertainty should be acknowledged so disclosure can become a progressive journey of learning. Let's allow ideas to unfold naturally before rushing to understand them.

Information as Evidence

Disclosure isn't limited by weak evidence. Radar data, pilot reports, sensor recordings, and repeated sightings have accumulated over many years. The real problem is elsewhere. The reported objects ignore inertia, cross barriers, and appear unexpectedly, which runs counter to our usual understanding of how things behave. When the evidence doesn't fit our framework, we often ignore it.

We must learn to interpret information without relying on outdated ideas about space, motion, and objects. Before the data can be understood, we need to develop the appropriate language for it.

The following section explores a deeper shift already underway in physics. This movement suggests that space, motion, and even objects may emerge from something fundamental: *relationships* and *information.*

Relationships Matter Most

At the most basic level, we understand that reality isn't composed of separate things. Space, time, and objects arise from the way information interacts. What truly matters isn't what things are, but *how they connect.* This may seem abstract, but modern physics has a solid mathematical foundation.

In holographic theories, the same physical system can be described in two equally valid ways. In the familiar view, particles move through space, encounter barriers, and obey the laws of gravity. In the more fundamental description, there is no space or gravity at all, only quantum information evolving on a lower-dimensional boundary. These are not different realities but two ways of describing the same one.

The boundary does not contain tiny particles or walls. Instead, it holds quantum information: states, correlations, and the rules governing how they change over time. Like a movie stored on a disc, this information doesn't resemble the final image but completely determines it. Space and motion are outcomes only when this information is translated into the familiar 3D picture.

DVD Player

This difference is important when we consider barriers and tunnelling. Typically, a particle moves toward a barrier and sometimes appears on the other side. In the boundary view, there is no barrier at all, only a change in how information evolves. Nothing passes through anything; the information updates according to fixed rules.

When we say that "boundary-encoded information evolves into a state where the particle appears on the other side," we are not describing magic or teleportation. The more detailed explanation doesn't involve a particle moving through space. Motion appears to occur only after the information is translated back into spacetime terms.

A simple analogy is a video game. The game's code doesn't track every step through a wall. The internal state changes, and the screen updates to show the character elsewhere. It may look mysterious, but the underlying description never required a path.

Video Game

This doesn't contradict physics. Energy, probability, and information remain conserved. Signals aren't travelling faster than light; the information changes in a lawful manner, producing the same results we observe in the familiar picture.

Measurement fits naturally within this perspective. You're not creating reality by measuring something; you're connecting to it through information. The change is correlation. This feels strange because our intuition is built around objects moving through space.

Reality might consist of relationships and information, rather than of objects moving through space. Space, motion, and barriers are what surface when that information is translated into the world we experience.

This idea hinges on the claim that *space is not fundamental*. It's often thought of as a backdrop that holds objects, but space actually arises from the relationships among entities at the quantum scale. It's not that space encloses entangled particles; instead, *entanglement gives rise to what we experience as space*. While this might sound unusual, it's a serious and respected idea in contemporary physics.

Network of Dots and Links (Entanglement Web)

Entanglement means two systems are linked so that you can't fully describe one without the other. Learning about one immediately reveals information about the other. This connection works regardless of distance and exists even before space is defined. That's why physicists think entanglement might be more basic than space itself.

Imagine building space using these connections. Each quantum system is like a point, and entanglement links them together. Strong links mean they are close, while weak links mean they are farther apart. From this view, distance comes from how strong the connections are. Space emerges from these connected patterns.

Physics supports this idea. In holographic theories, researchers have found that the amount of entanglement between regions determines the geometry of space. Change the entanglement, and the geometry changes accordingly. Without sufficient entanglement, smooth space cannot exist. In this sense, geometry is constructed from information.

Space comes from the power of connections

Imagine points connected by elastic bands, each representing entanglement. Tight bands make points feel close, while stretched bands make them feel far. The network's structure is what we perceive as space; when the bands are removed, space collapses.

This explains how space forms. Start with a quantum system that has no space, only information and connections. As these connections organize, ideas like "near" and "far" emerge. When the structure becomes complex, a smooth space appears that shows deeper relationships.

Physicists use diagrams called *tensor networks* to model this. These diagrams represent quantum states as points and entanglement as connections. Some of these networks naturally give rise to curved spaces and even extra dimensions. What appears to be a new spatial direction is a way of organizing entanglement, not a physical place you could travel to.

This idea also relates to the proposal known as "ER = EPR," which links entanglement to the structure of spacetime. In simple terms, entanglement acts as the glue holding space together. Reduce it, and the space stretches or breaks apart. Remove it entirely, and space disappears.

This means space isn't something objects sit in. It exists only as long as the relationships persist. Without those relationships, there is no real "here" or "there."

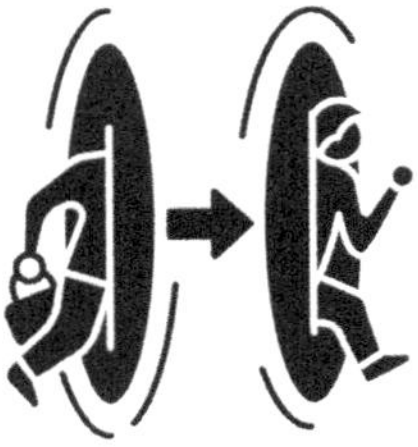

Quantum Tunnelling

This perspective also helps explain quantum tunnelling. When a particle appears on the other side of a barrier, it doesn't pass through in the traditional sense. At the most basic level, position isn't fundamental. What truly matters is the pattern of entanglement. If the relationships permit, the particle appears without requiring a physical path.

None of this suggests that physics becomes random or ruleless. Entanglement follows strict rules. Information is conserved, causality holds, and outcomes remain predictable. A helpful mental image is this: space is how our brains perceive stable patterns of entanglement, and motion is how we interpret changes in those patterns. At the deepest level, nothing truly moves through space; relationships are evolving.

All of these points point to an even more profound mystery: time. At its most fundamental level, the basic equations of physics do not include a built-in "now." Some theories describe the universe as timeless, even though we clearly experience change and flow.

How does time emerge from something that is timeless? At the fundamental level, the universe is timeless. What we perceive as time arises from how parts of the universe change relative to one another through entanglement and correlations. Space, motion, and time all originate from these basic patterns of interaction.

Measuring Time

Transition from Science to Society

Even as we begin to understand reality at its deepest levels, the universe doesn't reveal its secrets all at once. Disclosure, whether of UFO phenomena, hidden technologies, or paradigm-shifting truths, unfolds gradually, shaped by both human psychology and political structures.

Our minds evolved to interpret a stable and predictable world, but we are not equipped to handle sudden revelations of radically new realities. Evidence of non-human intelligence or hidden technologies provokes fear, denial, and cognitive dissonance.

Society needs time to process, absorb, and integrate these truths. People progress through stages: curiosity, skepticism, acceptance, and ultimately, integration. Each stage shapes how much information can be responsibly shared without overwhelming the public.

Political and institutional factors control how quickly information is released. Governments, military organizations, and bureaucracies often withhold information for strategic reasons, such as national security and social stability. Even if the public were psychologically prepared, those in power need to plan, assess risks, and carefully coordinate their communication.

History repeatedly reveals this pattern: nuclear technology, satellites, and other groundbreaking discoveries are gradually disclosed, allowing society to adapt without panicking. Disclosure succeeds only when public awareness and political frameworks are in sync. If one moves too quickly relative to the other, the process stalls or worse, backfires.

Global Society

This phased approach is intentional. Steady disclosure fosters trust, corrects false information, and helps cultural frameworks adapt to change. Sudden revelations can cause chaos, conspiracy theories, or social instability. By revealing truths gradually, they become integrated into the collective understanding, supporting better adaptation and comprehension.

Disclosure is a journey, not a single moment, and the universe may reveal its secrets, but humanity must be prepared to see them. Accurate understanding arises from the alignment of mind, society, and structure, not from shock. Only then can hidden truths become a stable part of our shared reality.

In recent years, disclosure has rarely occurred in a single dramatic moment. Instead, it has unfolded as a slow drip of information, allowing societies time to process and integrate new understanding.

The most prominent examples come from the United States, where military videos, declassified reports, and Congressional hearings have gradually confirmed the existence of unidentified aerial phenomena. Leaked and later officially released videos, such as *FLIR1*, *Gimbal*, and *GoFast*, allowed the public to confront the reality of unknown objects in the skies without making claims about their origin or intent.

At the same time, institutions such as the Pentagon's *All-domain Anomaly Resolution Office* (*AARO*) and Congressional hearings provided measured, systematic reporting that legitimized the phenomena without causing panic.

Canada has contributed as well, through efforts by the *Canadian Department of National Defence*, which in recent years has released previously classified reports on UFO sightings in Canadian airspace, particularly those involving military pilots.

Similarly, countries such as the *United Kingdom, France, and Chile* have released historical and contemporary reports, often through national archives or aviation authorities, indicating that slow-drip disclosure is a global trend. Each release is cautious and incremental: small enough to prevent mass panic yet significant enough to gradually reshape public understanding.

These international examples illustrate how disclosure is managed to balance readiness: the public's psychological readiness, governments' institutional readiness, and political strategy. Over time, these small revelations build familiarity, reduce shock, and allow a global audience to gradually accept that unexplained phenomena are real and worthy of serious study.

Recent Disclosure & UAP Reporting Timeline

Canada

- **2025:** The *Sky Canada Project* report was published, exploring how UAP sightings are gathered and suggesting improvements, with a full federal report expected by mid-year.

United States

- **2017–2020:** Leaked and subsequently officially acknowledged military videos (*FLIR1*, *Gimbal*, *GoFast*) brought mainstream attention to UAP phenomena.
- **2021–2024:** The U.S. Pentagon's All-domain *Anomaly Resolution Office* (*AARO*) was established and issued multi-year reports on UAP sightings and data collection efforts.
- **2024:** *AARO* published a historic review of decades of government UAP investigative programs.
- **2025:** A major documentary, *The Age of Disclosure*, premiered at a prominent film festival and was released worldwide, featuring testimony from government and intelligence officials and boosting public interest.
- **2025:** The U.S. National Archives expanded its public UAP record collection to increase transparency (ongoing).

France

- France's official UAP study group, *GEIPAN,* continues to operate within the national space agency and publicly releases case files and findings, treating UAP reporting as part of scientific aerospace monitoring.

Chile

- Chile's Section for the *Study of Anomalous Aerial Phenomena* (SEFAA) collects and scientifically investigates unexplained sightings, with ongoing public reports and analysis.

Russia / Soviet Union

- **Cold War era:** Thousands of UFO cases were investigated internally by military and scientific authorities, often classified as security concerns.
- **1989:** The *Voronezh UFO incident* received partial state media coverage; it remains one of the few publicized Soviet-era cases.
- **Present:** Independent organizations such as *Kosmopoisk* document sightings and anomalies, although no formal government disclosure program exists.

China

- **February 2023:** Authorities tracked an unidentified object over the Yellow Sea; it was reported as a security concern, but only limited details were released.
- Ongoing: Occasional military acknowledgments of sightings, while unofficial and social-media reports circulate widely; no systematic public disclosure program exists.

Other International Efforts

- The **United Kingdom** has declassified extensive sets of Ministry of Defence files from earlier UFO programs, making them accessible in public archives.

- Countries such as **Peru** and other **Latin American** nations produce annual case files and open reports, often in collaboration with scientific partners.

For decades, UFO films served as comforting fiction—symbols of Cold War anxieties or pop-culture fantasies. In the past decade, however, documentaries have surfaced that reflect an ongoing dialogue unfolding behind closed doors and in public hearings, reaching audiences now prepared to understand it.

Age of Disclosure, *Unacknowledged*, *The Phenomenon*, *Top Secret UFO Projects: Declassified*, and *Accidental Truth* mark shifts in cultural perception, echoing declassified intelligence, pilot testimony, and consciousness research. Together, they reveal a turning point where fiction and documentation converge. The dialogue centres on how to interpret the phenomenon and who controls the storyline.

These films emerged alongside societal changes. Congressional hearings removed UAPs from ridicule to serious scrutiny, while intelligence officials openly discussed sensor anomalies, non-aerodynamic crafts, and retrieval programs. The stigma diminished, and society became more receptive to a new narrative.

Collectively, the films highlight a public demanding transparency, institutions losing control of the narrative, coordinated whistleblowing, and a renewed debate over what humanity is permitted to know. Longstanding boundaries of science, myth, classified technology, and consciousness are breaking down.

Testimonies show credible military and scientific witnesses, reports of craft performing impossible maneuvers, and uncertain origins labelled as "non-conventional." Society's struggle to understand the unknown adds to the mystery. Filmmakers didn't create these stories; they gathered decades of scattered testimony, radar data, and research into clearer explanations. This clarity, more than any single sighting, is the main discovery.

Recurring elements include craft with no visible propulsion, extreme manoeuvrability, intelligent interactions with aircraft, connections to nuclear sites, and programs that disappear and re-emerge under new names. Secrecy is systemic: buried budgets, contractor buffers, internal uncertainty, and fears of public panic persist for decades.

Some films even explore the effects of consciousness, remote viewing, altered states, physiological responses, and sudden intuitive insights, suggesting that the phenomenon is as much psychological as technological. Each documentary underscores the same transformative truth.

Ancient parallels also emerge: myths of luminous objects, sky beings, cosmic messengers, and cycles of knowledge and renewal echo modern encounters. Private military contractors often possess more information than public agencies, and their access is protected by intellectual property law and NDAs.

Witnesses describe feelings of awe, disorientation, cognitive shock, fear, and a sense of importance. Across fields, from anthropology to cognitive science and religious studies, the evidence indicates that the phenomenon affects human consciousness as deeply as it challenges our physical understanding.

Part Two

HIDDEN INFASTRUCTURE

Chapter Four

Architecture of Secrecy

I keep thinking about what it must feel like to carry a massive secret you can't share. One that could change everything if it got out. After World War II, some people actually had that kind of knowledge. Scientists, soldiers, and officials were suddenly in possession of information powerful enough to reshape the world.

Over time, the mystery became something deeper. Small, clandestine groups, isolated offices, entire cities erased from maps, all shrouded in mystery. Everyone knew just enough to do their part, and all the information was divided into sections.

As a Gen X teenager, I noticed that, like most of us at the time, my parents were secretive. They would nod, smile, or change the subject, leaving us to piece together what we could, often guessing at what we couldn't see. We learned early that some things were meant to stay private and that you could only understand them by watching patiently.

I could only see a small corner of the picture and knew the rest was there, but for some reason, I wasn't in the loop. I had to trust the folks above me, the ones who could see the whole picture. That trust is the core of silence, turning it into the framework that defined things.

Secrecy is more than a safeguard for information; it's a shield as well. Public skepticism, disbelief, and ridicule form a powerful system that shapes our comfort levels. This system dictates what questions we dare to ask and what truths we dismiss. In essence, we are *policing ourselves*, suppressing our curiosity until it's less likely to seek the light.

Still, tiny leaks, cautious disclosures, and minor truths slip out. They don't reveal the full story, but they are enough to activate curiosity. They demonstrate that the world is larger, more mysterious, and more complex than we've been led to believe.

After World War II

Secrecy is often seen as a response to danger, yet history repeatedly shows the opposite: danger is usually defined by what institutions hide. After World War II, the United States changed.

The war showed that information could determine survival, that knowledge could be weaponized, and that a single technological surprise could change history. Similar insights were reached elsewhere. In Britain, France, the Soviet Union, and later China, wartime experience taught the same lesson: some knowledge was considered too deadly to be shared explicitly.

This realization led to a new culture of intelligence that viewed secrecy as the primary means of governance. The atomic bomb created a challenge that greatly affected countries around the world, highlighting the importance of secrecy for maintaining power and security.

Atom Bomb

Post-War Shock and Permanent Secrecy

For the first time, small groups of officials and scientists had information that could threaten civilization. Being open was once helpful, but became a problem, slowing quick action. Public talks just caused more doubt, delaying key decisions. As a result, real power went underground, making us more vulnerable than ever.

Across powerful nations during the Cold War, intelligence agencies expanded, classification systems proliferated, and entire domains of activity were concealed behind layers of clearance, code names, and deniability. The Soviet Union sealed entire cities off from public knowledge. Britain embedded lifelong silence in law through its Official Secrets Act.

France centralized strategic programs within narrow executive circles. China framed disclosure as a threat not only to security but also to social stability. What began everywhere as a wartime necessity hardened into a permanent policy.

In this environment, secrecy developed its own internal logic. Information was no longer shared to foster understanding but divided to keep it hidden. Compartmentalization ensured that no individual could see the full picture. Responsibility became unclear. Even those working within these systems often found it difficult to distinguish what they knew from what they weren't supposed to ask. The framework remained secure because it was incomplete.

Confidential

In this global context, MJ-12 became a symbolic idea; the notion of a small, insulated group handling extraordinary information aligns with postwar intelligence practices. MJ-12 existed as an entity in which information was deemed too destabilizing for democratic or civilian systems; secrecy justified itself.

Over time, secrecy moved from protecting information to defending itself. Sharing secrets was too risky because of decades of denial and misdirection. Each generation inherits these secrets without understanding their full story, and is only warned not to reveal them.

Institutional Gatekeeping

Compartmentalization became the central principle of modern times. Information was released only on a "need-to-know" basis. Each individual received only a small portion of the technical details, with context lacking, goals devoid of background, and results presented without explanation. No one person or office had the full picture.

This effectively protected sensitive material from exposure, but it also made meaningful interpretation nearly impossible.

Insight into the wider system weakened, ethical oversight eroded, and errors could go uncorrected.

The Manhattan Project is a clear example. Tens of thousands of scientists, engineers, and workers contributed to its success, yet most were unaware they were building an atomic weapon. Laboratories were isolated from one another, research was divided by function, and even senior officials were blind to the full picture.

Similar practices appeared in Soviet nuclear and aerospace programs, where entire "closed cities" were omitted from maps and personnel were kept in lifelong silos of knowledge. During the Cold War, *NATO* black programs used similar formats, distributing classified components across allied countries while ensuring that no single partner could reconstruct the whole.

Over time, this compartmentalization altered how truth operated within secret systems. The system relied on obedience, prioritizing procedure over judgment and understanding.

MJ-12: The Hidden Structure of Authority

MJ-12, short for *Majestic 12*, is the name given to an alleged secret committee said to have been formed by the U.S. government in the late 1940s to manage information about unidentified flying objects (UFOs) and potential non-human intelligence.

According to documents that surfaced in the 1980s, MJ-12 was composed of senior military leaders, scientists, and intelligence officials who operated outside normal oversight and reported directly to the highest levels of government. Its purported purpose was to investigate recovered materials, assess national security risks, and control public disclosure.

The existence of MJ-12 has never been officially confirmed,

and many historians and analysts argue that the documents are either fabricated or partially altered. However, others note that a highly compartmentalized group closely aligns with known post World War II intelligence practices.

Across the globe, postwar powers managed knowledge that was too sensitive, destabilizing, or dangerous for governance. Britain, France, and later NATO countries relied on small, highly restricted committees to oversee nuclear, intelligence, and black programs, preventing any single official or nation from reconstructing the full picture. Even today, China maintains highly segmented military and technological programs, isolating personnel from the broader operational and political context.

Splitting knowledge, spreading out responsibility, and often ignoring ethical or strategic concerns is about controlling what people understand.

MJ-12, whether real, symbolic, or imagined, illustrates how small secret groups form to control information and shape how people perceive the truth.

Uncontainable Secrecy

As secrecy became a self-sustaining system, shielded from scrutiny, what was once meant to protect society began to influence it in inconspicuous ways.

Public skepticism transitioned to a means of manipulation. The more information is withheld from the public, the greater the incredulity. Ridicule replaced rational rebuttal; skepticism was socially embedded, and challenging the official story risks censure, mockery, or removal.

Cultural stigma, manifested as fear of being branded credulous or paranoid, strongly enforces compliance. This pathology has significant effects when disclosure finally occurs. Information leaks or official acknowledgments are often inconsistent or incomplete.

Witnesses and insiders who choose to come forward are limited by what they were allowed to know, and their testimonies are influenced by decades of secrecy. As a result, the public encounters confusing, emotionally intense information because the system designed to safeguard it has crippled understanding at all levels.

Puzzling Disclosure

UAP Disclosure and Government Transparency

In the U.S., there has been a surge in official hearings, media coverage, and documentaries promoting transparency around *Unidentified Aerial Phenomena* (*UAP*). Recent congressional hearings featured an unauthorized release of video of a missile targeting an unidentified object off Yemen, which attracted significant public interest and debate, especially because it was *leaked*.

A prominent documentary, *The Age of Disclosure,* features former officials, including those associated with government UAP programs, who allege that information has been systematically withheld from the public. Some contributors contend that intentional disinformation campaigns have reinforced this secrecy and that the stigma surrounding the topic has hindered scientific research and public awareness.

Meanwhile, political discussions continue over whether

task forces and legislation should require the declassification of UAP-related records, with bipartisan calls in Congress for greater transparency.

Public reaction is widespread and mixed: from eager anticipation and the circulation of whistleblower stories to strong skepticism and media disputes over credibility. For many, the conversation isn't only about *what* is revealed, but also about how, why, and by whom information leaks into the public domain.

Across these cases, public response is often polarized. Some see leaked or declassified material as a reclaiming of collective knowledge long withheld by powerful institutions. Others view these exposures with skepticism, alleging misinformation or questioning the motives of leakers and commentators.

From Newsrooms to Podcasts

In digital communities such as online forums and social media, discussions about disclosure trends range from organized calls for government transparency to speculative narratives about suppressed truths. These platforms amplify both *trust and distrust* in official channels, showing that disclosure in the 21st century reshapes public discourse about credibility, authority, and belief.

In today's era of disclosure, media platforms and podcasts have become key spaces where competing narratives about secrecy, UAPs, and hidden programs unfold. One example is the podcast *Voices of Disclosure: The UFO Experts*, which examines government secrecy, extraterrestrial encounters, and declassified documents, featuring commentary from leading researchers such as *Richard Dolan* and *Dr. Steven Greer*. The show combines witness testimony and historical cases with discussions of official records, demonstrating that disclosure debates often involve both interpretation and evidence.

Stephen Greer also hosts his own podcast, *The Dr. Steven Greer Podcast*, where he discusses emerging topics related to UAPs, covert investigations, consciousness, and the broader effort toward disclosure, as he defines it. Episodes in 2025 have explored former military encounters and the relationship between classified programs and public awareness, demonstrating that Greer continues to influence the discourse on disclosure.

Podcasts

The Total Disclosure Podcast with *Ty Roberts* examines UFOs, paranormal phenomena, and government cover-ups by interviewing researchers, scientists, and former insiders. These podcasts aim to peel back "layers of secrecy" and reach audiences who feel that traditional channels have suppressed specific topics.

Another example is the *Universal Disclosure Podcast*, which blends UAP discussions with an exploration of how these phenomena connect to spirituality and personal belief systems, illustrating how disclosure topics often extend into broader cultural and existential questions.

This media ecosystem insinuates that disclosure today is not only about information released by governments but also about how narratives spread, evolve, and compete. Some audiences seek scientific foundations and evidence-based reporting, while others are drawn to experiential or spiritual interpretations.

We now navigate a landscape in which public understanding of disclosure is shaped by the voices and formats that gain prominence, from mainstream documentaries and congressional testimony to independent podcasts that blend speculation, testimony, and analysis.

China–Moon–helium-3

Until now, space has mainly been seen as a realm for observation, using satellites and sensors. Yet, observation gradually turns into interest, and interest into purpose. When a region can be measured, it becomes assessable; when it's accessible, it gains significance. The moon represents the point at which space shifts from something we observe to something more personal and familiar.

Publicly, lunar exploration is still framed in scientific and cooperative terms. Privately, it has begun to resemble entirely different activities: early-stage infrastructure planning. No nation exemplifies this shift more clearly than China.

Although commercial lunar mining has not yet begun, China's lunar program (launched in 2004) has focused on resource assessment. Mapping missions, sample returns, and technology demonstrations form a sequence that seems more like a decades-long feasibility study than pure exploration.

China's early *Chang'e* missions focused on elemental mapping, determining the chemical composition of the lunar surface and identifying materials with long-term value. Later missions progressed from observation to retrieval.

The return of lunar samples in 2020 marked the first such recovery in more than forty years. Analysis of those samples revealed a previously unknown lunar mineral containing helium-3. Helium-3 is an isotope long theorized to be a cornerstone of future fusion energy.

More recent missions have expanded this effort to the Moon's far side, targeting regions believed to preserve some of the Solar System's oldest material and possibly higher concentrations of strategically valuable elements.

The importance of these missions lies in their main goal. Future trips are planned to test how to use lunar resources on-site and to find ways to extract, process, and use materials from the Moon without returning them to Earth.

Technologies such as autonomous excavation, soil processing, and even 3D printing using lunar regolith (the powdery, rocky debris covering the lunar surface) signal a transition from a temporary presence to sustained operations. Plans for a permanent lunar research station in the coming decades further support this trend.

Helium-3, on Earth, is found only in trace amounts. On the Moon, billions of years of solar wind exposure have embedded it deep within the upper layers of the lunar soil. In theory, helium-3 could enable fusion reactions that generate large amounts of energy with very little radioactive waste.

In practice, retrieving it would require processing vast amounts of regolith in an environment hostile to both machinery and human life. The technology needed to exploit it at scale doesn't yet exist, and controlled helium-3 fusion remains an unresolved challenge.

Lunar resources will be exploited, but who will be positioned to do so first once the technical barriers are overcome? *The Outer Space Treaty* prohibits national ownership of celestial bodies, but it does not clearly address ownership of extracted materials, creating a legal grey area that benefits those already operating on-site.

What's often overlooked is how quickly exploration can turn into an advantage and how future dominance is set in motion. In a world where space is no longer seen as neutral, access to it is power, and the biggest moves are often made years before the public realizes a game is underway.

Chapter Five
Shadow Matrix

The first time I saw a convincing deepfake, I knew we all were going to experience things differently going forward.

If you're reading this, you've likely noticed it too. Uncertainty about what to trust is growing. News clips often look staged, and social media posts go viral before they're verified. Images stay vivid in our minds even as doubts arise. Quick, repeated exposure before we can analyze them.

Power has always preferred distance and deniability, and deepfakes offer both. This isn't by accident.

Deepfakes

Deepfakes represent a significant transition in how power operates. Generating convincing false realities is now accessible to more than just governments or major studios. Voices can be imitated, faces reconstructed, and events fabricated using tools that fit on a laptop. When reality can be changed, the power over truth belongs to those who possess the tools.

Deepfakes work because of how they look. A fake speech or scene can seem real and doesn't need any background if it looks authentic. By the time people realize it's fake, it may have already affected trust, reputation, or public opinion.

This technology provides a considerable edge for black projects and private power because confusion often proves more effective than persuasion. The public is uncertain about whom or what to trust and is hesitant to act. In such situations, there is no need to hide power.

The deeper issue is why it's hard for people to prove authenticity. Genuine evidence can be dismissed as fake, and witnesses can be discredited. Unfortunately, private interests operate with little to no regulation.

AI Deepfakes

Defence Contractors as Gatekeepers

History shows that defence contractors tend to expand beyond their original mandate when oversight weakens. This happens across political systems, in multiple countries and eras. When military functions are outsourced, decision-making is carried out by the contractor rather than the nation, especially when technical expertise is held in private hands.

United States

During the wars in Iraq and Afghanistan, companies such

as *the Halliburton* subsidiary *KBR* were awarded extensive logistics and reconstruction contracts with minimal supervision. Subsequent investigations uncovered inflated costs, unfinished work, and instances in which contractors effectively managed supply chains that the military could not independently verify.

At the same time, major defence companies working in intelligence used special systems that congressional staff couldn't fully review because they lacked the necessary clearance or technical access. As a result, oversight depended on reports from the contractors.

Long-running programs tied to aerospace, signals intelligence, and weapons development have repeatedly dodged congressional visibility. Contractors have drafted technical requirements, assessed their own performance, and maintained control of systems that the government couldn't audit independently.

United Kingdom

Similar patterns emerged in the United Kingdom, where private companies assumed roles in intelligence analysis, logistics, and cybersecurity. This erases the boundary between support and operational control.

Private firms such as *G4S* and *Serco* assumed security, detention, and military roles beyond the scope of standard contracts. In several instances, companies were found to have billed for services they didn't provide, including staffing prisons and monitoring detainees. Contracts were repeatedly renewed because there were no in-house options, showing how reliance on private companies limited opportunities for correction.

France

French defence contractors continued supplying arms and surveillance equipment to foreign governments despite periods of political instability or human rights concerns. Parliamentary inquiries later revealed that export controls were often applied after contracts were signed, not before, allowing commercial commitments to override policy shifts. Once embedded, these agreements proved difficult to reverse without diplomatic or financial fallout.

Germany

German arms manufacturers were implicated in unauthorized weapons exports, including small arms that appeared in conflict zones despite export restrictions. Internal compliance warnings were documented years before public exposure. Enforcement followed only after external investigations, suggesting oversight mechanisms lagged behind contractor activity.

Russia

Private military companies, such as the *Wagner Group*, have operated alongside official forces in Ukraine, Syria, and parts of Africa. Although technically independent, these groups relied on state-provided equipment, logistics, and intelligence support. Their ambiguous status enabled the government to disclaim responsibility for actions while still securing strategic benefits, leaving no clear path to legal or political accountability.

China

Chinese technology and defence companies developed

surveillance systems initially justified on national security grounds, which were later used in civilian governance and exported internationally. Contracts between local governments and private firms often lacked transparency, and capabilities frequently expanded after deployment. The overlap between commercial development and state security goals made it difficult to separate public policy from corporate initiatives.

Israel

Israeli defence and surveillance companies tested technologies in conflict zones before promoting them internationally. Crowd-control systems, facial recognition software, and monitoring tools later appeared in civilian policing and border control abroad. Oversight focused on export approvals, limiting accountability once the systems left state control.

India

Private defence and technology companies contracted to provide biometric identity and surveillance infrastructure gained long-term access to sensitive data systems. Subsequent audits raised concerns about data security and mission creep, as tools designed for security or welfare verification expanded into broader population monitoring without clear legislative debate.

South Africa

Private security firms, some with military origins, became deeply embedded in domestic security functions after apartheid ended. Weak regulation allowed companies to influence policing practices and intelligence sharing. Public and private authorities were later difficult to differentiate.

Latin America

Several countries procured surveillance and interception systems from foreign defence contractors to combat organized crime. In many cases, these tools were later used against journalists, political opponents, and civil society groups. Investigations found that contracts were often signed under secrecy provisions that limited judicial or legislative review.

Canada

Defence and security contracting has expanded steadily through long-term relationships between the federal government and a small number of private firms, particularly in aerospace, intelligence support, and surveillance technology. Companies involved in military communications, satellite systems, and data analysis have operated with limited public visibility under national security exemptions, leaving parliamentary oversight largely dependent on executive reporting.

One example is the outsourcing of intelligence and data analysis functions to private contractors who work alongside the *Canadian Security Intelligence Service* (CSIS) and the *Communications Security Establishment* (CSE). These contractors often possess technical expertise that the government does not maintain internally, creating reliance and complicating independent review. In several cases, oversight bodies acknowledged difficulty assessing systems they could not fully access or replicate.

Canada's role in arms exports has also raised concerns. Canadian-made light armoured vehicles sold to Saudi Arabia continued to be delivered despite public controversy and legal challenges over human rights. Once contracts were signed, commercial and diplomatic pressures limited the government's ability to alter or halt deliveries, demonstrating how contractual commitments can thereafter constrain policy.

Surveillance technology offers another example. Canadian firms have developed and exported interception and monitoring tools marketed for lawful access and public safety. Investigations later found similar technologies used abroad for political surveillance and repression. Export controls focused on formal licensing instead of downstream use, leaving accountability diffuse once systems left Canadian jurisdiction.

Domestically, public-private partnerships in defence, such as long-term aircraft maintenance and naval support contracts, have shared operational knowledge with private companies but have also made the government dependent on contractors for cost, readiness, and risk assessments. Reports from the Auditor General have repeatedly highlighted cost overruns, delays, and limited control once contracts are in place, with few options other than cancellation.

In these cases, the main problem is that authority is exercised through contracts.

Over time, defence contractors begin to shape policy, wielding influence that is difficult to trace and even harder to challenge.

The Black Box Problem

A helpful way to grasp the black box problem is to picture a single bee in a hive. You could watch it for months, recording every move and noting how it collects nectar, cares for larvae, or communicates through the waggle dance. Every detail could be documented.

Yet no matter how long you study a single bee, you would never understand how the hive organizes its resources, decides when to swarm, or maintains its internal temperature. The hive's behaviour emerges from interactions among thousands of bees. The individual bee knows nothing of the bigger picture, and no observer of a single bee can predict the colony's collective decisions.

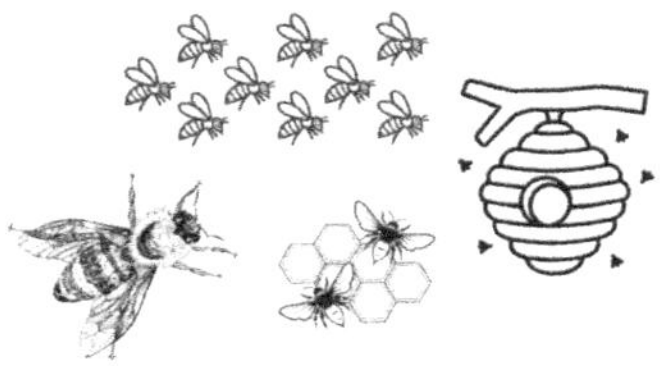

Bees

Modern defence and security systems operate in a similar manner. Engineers, analysts, and operators each see only a part of the system. Components are developed, trained, or deployed in isolation, often without understanding how they interact downstream.

Even senior managers rely on dashboards and summaries, creating a complex system whose behaviour arises from interactions across layers, making understanding and accountability difficult.

Hidden Tech

History offers examples of technologies developed in secret that eventually become public knowledge, often long after their initial deployment. The *U-2 spy plane,* tested in near-total secrecy in the 1950s, baffled both domestic and international observers until its details were declassified. Its capabilities, altitude, speed, and surveillance range seemed almost impossible at the time, prompting reports of "unreal" aircraft.

Once the program was disclosed, what had seemed unusual

fell into place: the technology was there, but no one outside the program could see it.

Similarly, stealth aircraft in the 1970s and 1980s operated under strict classification. Radar operators and pilots sometimes recorded objects that moved or vanished in ways they could not explain. Publicly, these sightings were dismissed as sensor errors or optical illusions. Internally, they offered early glimpses of experimental systems meant to manipulate detectability.

The pattern repeated with satellites, communications, and early laser or electronic warfare tests, producing partial and conflicting data observed by those not cleared for the program.

Even in more recent decades, technologies such as drones and high-speed unmanned aircraft suddenly appeared in public or media reports. Programs developed under private contracts or by defence agencies often produced sensor anomalies before they were officially recognized.

Analysts working on this faced a tricky challenge: the behaviour seemed impossible at first, but once the context was understood, the apparent contradictions were cleared up by sophisticated, well-organized human-made systems.

These examples highlight an essential point for UAP research. Apparent anomalies do not necessarily indicate a non-human origin. Instead, they might result from highly advanced, secret technology interacting with sensors in unexpected ways.

Reverse Engineering

Several individuals have publicly claimed firsthand involvement in programs studying anomalous technology. While the ultimate truth of these accounts remains disputed, they reveal compartmentalized work, limited context, and research on systems that appear disconnected from conventional development paths.

Bob Lazar (United States)

Lazar claims that in the late 1980s, he worked at a site near Area 51, studying propulsion systems of recovered craft. He said his role focused narrowly on a reactor-like power source that generated gravity effects, with no access to the craft as a whole. Lazar described working with others similarly isolated and with materials that were extremely lightweight and resilient.

While his account remains controversial, the compartmentalization he describes mirrors that of known black programs, such as early stealth aircraft development, in which engineers had only partial access to subsystems and breakthroughs were distributed across teams to protect secrecy.

Philip J. Corso (United States)

Colonel Corso claimed in *The Day After Roswell* that he oversaw the distribution of recovered technologies to private defence contractors, seeding them into research under conventional pretexts. He attributed advances in integrated circuits, night vision, and materials to this process.

Whether these claims are accurate or not, they resemble the historical practice of reverse-engineering foreign or captured technology, as seen in postwar U.S. jet propulsion and rocketry programs, where fragments of enemy hardware informed incremental progress in domestic laboratories.

Eric Davis (United States)

Physicist Eric Davis has stated that recovered off-world materials were studied under extreme restrictions, often yielding inconclusive results. While he avoids claiming breakthroughs, the environment he describes parallels declassified programs such as the *Advanced Research Projects Agency (ARPA)* studies on exotic materials and aerospace prototypes, where iterative testing preceded broader understanding.

David Grusch (United States)

Grusch, a former intelligence officer, has claimed knowledge of long-standing crash-retrieval and reverse-engineering programs involving both government and private contractors. His emphasis is structural: programs operated outside oversight, with information tightly siloed. This reflects patterns confirmed in declassified projects such as the U-2 and SR-71 programs, where only small teams had access to the entire system and operational knowledge was tightly controlled.

International Parallels

Similar patterns appear outside the U.S. In the Soviet Union, Cold War-era researchers sometimes received anomalous materials with no provenance and were instructed to characterize their properties in isolation. Declassified French and UK defence projects also show compartmentalized study of novel aerospace incidents. In each case, insiders worked on pieces without understanding the whole system—a pattern that aligns directly with the experiences described by Lazar, Corso, Davis, and Grusch.

The overlap between insider claims and declassified programs shows that secrecy, compartmentalization, and incremental understanding are consistent features of advanced defence and aerospace development, regardless of the extraordinary or controversial nature of the specific programs described.

Men in Black (MIB)

Reports of so-called *Men in Black* have become a persistent feature of modern UFO lore, between rumour, psychology, and alleged covert activity.

Individuals who claim to have witnessed unusual aerial phenomena sometimes describe a second event that occurs shortly afterward: an unexpected visit or contact with figures whose apparent aim is to discourage further discussion of what was observed.

In many reports, these encounters are very quick. Witnesses report being contacted within hours, sometimes sooner, before they have spoken publicly or confided in close friends. The visitors are often described as arriving in large, dark vehicles and wearing formal black clothing that seems out of place or old-fashioned.

Their behaviour often stands out more than their appearance. Movements are described as stiff or mechanical, speech patterns as oddly dated or rehearsed, and emotional expression as flat or absent. They provide reassurance or authority in a typical way, leaving an impression of something performative.

Claims about identities vary. Some witnesses report being shown identification that appears official at first glance but doesn't hold up under closer inspection. Others describe the figures as claiming vague ties to the government, while still others recall them posing as journalists, salespeople, or military personnel. This inconsistency only heightens unease, as these encounters seem intended to convey authority without clearly establishing it.

Accounts vary significantly in their details. Some involve multiple in-person visitors, while others are just a phone call or a brief conversation. In some cases, witnesses recall direct warnings to stop investigating or discussing their experience. In other cases, the interaction is more ambiguous, marked by probing questions and cryptic remarks.

A common element in many stories is the idea that the visitors knew personal information the witness hadn't publicly disclosed, which heightened the feeling of being watched.

What really gives these stories their lasting power is their shared atmosphere. Across decades and cultures, the *Men in Black* are remembered more for how they make people feel than for what they say: disoriented and intimidatingly pressured into silence.

Whether viewed as agents of secrecy, as psychological projections driven by stress and fear, or as evolving folklore rooted in Cold War anxieties, the phenomenon persists because it reflects a deeper concern that extraordinary knowledge is managed, contained, and strongly discouraged from coming to light.

Chapter Six
Mind Games

If you've ever hesitated to talk about something strange you've experienced...

If you've ever come across a credible account and immediately looked for a reason *not* to trust it.

If you've ever thought, "I don't know what's real anymore, so why bother?"

Then this chapter is already about you.

* * *

At some point, many of us realized that *nothing sounds true anymore*. Not just the strange things or fringe ideas. Everything. Official statements, breaking news, expert opinions, viral footage, even our own instincts.

You don't need to lie to control people. You don't even need to hide the truth. All you have to do is make confidence seem naïve. Once people are unsure, they disengage and retreat. They learn to laugh things off before they're laughed *at*.

I started studying psychological operations because I recognized how involuntary our reactions are, especially to unexplained phenomena like extraterrestrials, consciousness, strange events, and topics that challenge science. Usually, our first responses are humour, dismissal, or embarrassment.

If truth always seems out of reach, it's worth asking whether that distance serves someone.

Psy-Ops, Misinformation & Manufactured Doubt

Let's begin with a brutal truth: control doesn't need to convince people that something is false; it only needs to make them have a doubt. This idea is central to psychological operations (psy-ops), a field that has evolved over more than a century.

Governments, militaries, and intelligence agencies have realized that shaping perception can be more effective than controlling facts, long before social media accelerated confusion. When people cannot decide what to believe, they stop questioning.

This conditioning is most apparent in how society has been trained to respond to extraterrestrial, anomalous, or supernatural topics. It isn't about inventing lies from nothing, but about adding so much noise to real signals that they lose their meaning.

Playbook with a Proven Record

Weaponized doubt has a long history, notably in industry. In the mid-20th century, tobacco companies encountered growing proof connecting smoking to cancer. Instead of challenging the science, they sought to delay agreement by claiming it was still "unresolved." Internal documents clearly outlined their approach: the deliberate creation of *doubt was the goal.*

Cigarrettes

Cold War psychological strategists quickly learned this lesson. Later, the same reasoning appeared in national security settings.

The default explanation was vague, suggesting weather balloons, sensor errors, or a lack of information. Over time, this fostered a cultural reflex: if something sounds strange, it is probably nonsense. Uncertainty has been deliberately sustained.

Ridicule proved even more effective than doubt. Censorship sparks curiosity, but mockery extinguishes it. Cultural shorthand such as "little green men," tinfoil hats, and X-Files music did not arise by accident.

Intelligence agencies openly admitted to using ridicule to deter witnesses, discredit researchers, and prevent serious investigations from gaining institutional credibility. Once ridicule became established, enforcement wasn't necessary.

Ridicule

Scientists have found that engaging with anomalous data can derail careers. Pilots discovered that reporting unusual encounters could damage their reputations. Every day, people realized that speaking openly about strange experiences risked being labelled irrational or worse, as schizophrenic.

Another key tactic was saturation. Authentic reports circulated alongside hoaxes, sensationalist personalities, fabricated stories, and wildly contradictory claims. Serious Pentagon briefings coexisted with alien autopsies, psychic channellers, and CGI videos.

To the public, everything was lumped into one category, "that woo-woo stuff." This attitude didn't require rational debunking. Faced with overwhelming contradictions, people tended to dismiss them regardless of the evidence because rejecting the topic entirely felt safer.

Over time, this strategy had a secondary effect: the erosion of institutional trust. Repeated denials followed by partial admissions taught the public that official narratives were provisional and often unaccountable.

When governments eventually acknowledged unidentified aerial phenomena or admitted that data had been withheld, the response was *fatigue.*

A *disengaged public* is easier to manage than one that persistently questions technology, reality, or consciousness.

Extraterrestrial and supernatural themes are especially vulnerable because they challenge what we consider real. Topics such as non-human intelligence, non-biological sentience, and unknown forces intersect with religion, philosophy, geopolitics, and power. Such questions threaten existing hierarchies.

Case Studies

In November 2004, radar operators aboard the *USS Princeton* detected multiple anomalous objects off the Southern California coast. These objects were tracked for several days, appearing at altitudes above 80,000 feet, descending rapidly to near sea level, and exceeding the capabilities of aircraft. Radar personnel repeatedly recalibrated their systems, suspecting a malfunction, but the anomalies persisted.

When Navy pilots were dispatched, trained aviators visually confirmed the radar data. Commander *David Fravor* later described an object with no visible propulsion or wings and capable of instantaneous acceleration. Overlapping data from infrared footage and targeting systems corroborated it.

The official response was that it was sensor glitches, atmospheric effects, and unidentified drones. Without reconciling these explanations with the actual observations, the case remained unresolved.

* * *

A similar event took place in 1967 at *Malmstrom Air Force Base* in Montana, when personnel observed a luminous object hovering near nuclear missile silos as several intercontinental ballistic missiles went offline simultaneously. Launch officers with high-level clearance recorded the incident.

Follow-up investigations acknowledged the missile failures but officially separated them from the sighting. No public analysis of the cause was provided. The unexplained aspects were dismissed, and the incident was presented as an unrelated technical failure.

* * *

In the 1970s, U.S. intelligence agencies funded remote-viewing experiments through programs later known as *Stargate*. Researchers from various fields tested whether non-local perception had practical value. Results varied, but were not null. Internal reports confirmed statistically significant results under controlled conditions.

When the program was declassified in the 1990s, public messaging highlighted methodological flaws and ambiguity, portraying the research as a failure despite internal disagreements. No effort was made to situate the findings within the broader scientific inquiry. The public verdict was simple: it "didn't work," so it was senseless.

Proposed possibilities remain unresolved, and as time passed, interest diminished. Over the decades, this repetition conditioned the public. This didn't develop organically; it was learned through repeated exposure to institutional non-answers masquerading as closure.

That may be the most enduring result of manufactured doubt. It persuades people to dismiss the extraordinary as futile to deal with.

Project Blue Book

The U.S. Air Force's *Project Blue Book* (1952–1969) is often cited as evidence that UFOs were thoroughly investigated and regarded as insignificant. The reality is more complex and revealing.

While the Blue Book publicly concluded that most sightings were misidentifications or natural phenomena, internal assessments indicated otherwise.

The *Robertson Panel* (1953), convened by the CIA, explicitly recommended reducing public interest in UFOs by:

- educational debunking
- media partnerships
- ridicule of witnesses

Widespread interest in UFOs was seen as a potential national security threat, capable of overwhelming communication systems or being exploited by adversaries during moments of panic.

Governments repeatedly change their positions, denying interest in UFOs, then admitting secret programs, dismissing sightings, and finally confirming unknown craft. By the time the U.S. Department of Defence recognized UAPs as real but unexplained in the late 2010s, the public response was uninvolved.

Monitoring Public Reaction

The modern discussion of UAPs didn't begin with Pentagon press releases or congressional hearings. It began decades earlier, during the Cold War information climate, when perception was treated as a strategic tool. To understand why official recognition of UAPs in the 21st century was met with indifference, it's essential to examine how public certainty was deliberately eroded long before disclosure.

Psychological operations are often seen as deception, but their main purpose is to corrode trust. When confidence in knowledge drops, people naturally comply or disengage. UAPs were not treated as aerospace technology or alien intelligence.

By the time official acknowledgment arrived decades later, the psychological groundwork had already been laid. When the Department of Defence confirmed the existence of UAPs and admitted that some defied current explanation, the response was unenthusiastic. The dominant reaction was resignation.

UAP disclosure journalism now operates within this legacy. The central challenge for disclosure reporting is whether public trust, once strategically stifled, can be rebuilt sufficiently to enable meaningful engagement.

Countering the Mind Games

If doubt is used as a weapon, then being able to tell what's true is the best defence. The aim is to get back the ability to judge claims without losing curiosity. This is harder than it seems because the systems that broke trust were designed to make judgment risky.

The first step is recognizing that skepticism has been weaponized. Healthy skepticism questions and assesses evidence, whereas weaponized skepticism dismisses inquiries before they can be raised. It portrays engagement as foolish and curiosity as gullibility. In this environment, the phrase "extraordinary claims require extraordinary evidence" is often used.

To address this, we must distinguish between uncertainty and ignorance. Uncertainty indicates that the data is incomplete, whereas ignorance signifies a refusal to engage with available data. Many public debates intentionally cloud the distinction between these two states, treating an unresolved issue as disproven.

A disciplined response helps avoid shortcuts by allowing time to consider unanswered questions rather than rushing to resolve them.

Open-Mindedness VS Blind Faith

One of the most effective tactics used to discredit inquiry is to equate openness with gullibility. Anyone willing to consider unconventional explanations is portrayed as disconnected from reality.

Open-mindedness is the willingness to revise conclusions in light of new data. Credulity, by contrast, is belief without standards. The two are opposites.

Countering psychological conditioning involves reclaiming this distinction internally. It means being able to say, *"I don't know yet, but I'm paying attention,"* without feeling compelled to defend or immediately demonstrate certainty. It also means tolerating the discomfort of holding multiple hypotheses without rushing to resolve them.

Institutions benefit when people demand quick answers, yet nature, science, and complex systems rarely provide them.

Pattern Recognition

Another vulnerability exploited by misinformation is the human tendency to see patterns. Pattern recognition is vital to intelligence, but when trust breaks down, it can devolve into *pattern fixation.* In an information-rich environment, every coincidence can seem meaningful, and every inconsistency can appear intentional.

A helpful way to approach questions isn't just to ask *"Is this true or false?"* but to consider *"What pattern does this resemble, and how frequently has it appeared before?"* When mechanisms such as ridicule, ambiguity, compartmentalization, and delayed acknowledgment recur across cases, fields, and countries, this suggests a deliberate intent, even if the facts are uncertain.

Noise

One reason people withdraw from complex topics is *cognitive overload.* People disengage when everything seems contaminated by official statements, alternative media, experts, and influencers. This is intentional. Saturation aims to make an independent evaluation feel impossible.

Truth isn't just one source; it comes from overlaps. Always compare evidence to verify.

- data gathered from various sensor systems
- reluctant admissions versus voluntary claims
- statements *before and after* declassification
- consistency over time, not confidence in tone
- independent witnesses without shared incentives

Importantly, this approach doesn't need full access to classified information. It depends on the experiences of professionals over the years who have reported similar limitations, described comparable forms of compartmentalization, and faced similar institutional responses; these shared experiences provide valuable insights.

Authority should never be the final point of inquiry based on a single piece of data among many others.

Dare to Speak

One subtle effect of ridicule-based conditioning is self-censorship. People learn not only what to say but also what *not* to say aloud. Over time, internal curiosity is suppressed before it can be expressed.

Countering this begins privately, by allowing oneself to take a question seriously without needing immediate validation. It also means resisting the urge to prematurely mock one's own curiosity.

This is especially crucial in the context of incredible experiences. Many individuals have encountered phenomena that don't align with accepted ideas. The cultural response has trained them to reinterpret or downplay these events before others do. This internal editing is one of the most effective forms of control because it doesn't require enforcement.

Restoring confidence starts by recognizing mysterious experiences without rushing to explain them.

Foolish Certainty

It may seem counterintuitive, but the solution to manufactured doubt is *adaptive reasoning*. Science advances in this way, as do intelligence analysis and philosophy.

If manufactured doubt led people to believe that asking questions is pointless, countering it means making questioning meaningful again, even when the answers aren't perfect.

The main danger to systems that rely on managed perception is people's ongoing interest, which is hard to regain once lost.

Part Three

PRIMORDIAL MEMORY

Chapter Seven

Mythic Archives

In 2013, when I first created a Facebook account, one of the profile fields asked for religion. I remember staring at it, unsure what to input. I didn't fit neatly into any established category, yet I knew I wasn't an atheist. I believed consciousness mattered in unexplained ways, though I had no clear word for it.

That word came to me through a thorough internet search as *panpsychism*, the idea that consciousness is not an accidental by-product of matter but a fundamental aspect of reality.

When I saw it, I instantly understood. I finally found a name for something I'd felt for years: that mind and matter aren't separate, and that many old traditions might be simple misunderstandings. Our ancestors had different ways of expressing the same basic truth.

The idea is that consciousness isn't an accidental result of complex minds but a basic part of reality. Mind, in some form, might be built into matter, like charge or gravity, and is older than human consciousness.

The clear separation between mind and matter, observer

and observed, living and nonliving, wasn't solved right away by panpsychism, but it helped smooth out persistent differences.

It provided a way to see consciousness as more than something that suddenly appears at a certain level of biological complexity. Instead, life is organized, just as temperature arises from motion and sound from vibrations.

Ancient texts didn't read as naive mythology; they read as primitive, symbolic, and culturally isolated attempts to describe encounters with an intelligent universe, using the only language available at the time.

Gods. Angels. Sky beings. Watchers. Messengers. Wheels within wheels. Voices from fire, clouds, or light. For centuries, we've regarded these accounts as metaphors, allegories, or superstitious stories meant to explain the unknown or control populations. And sometimes, without doubt, they were.

I suspect that some of these narratives were meant as actual records.

Oral traditions were humanity's first way of preserving memories. They shared stories, repeated important details, used symbols, and performed rituals to pass information down through generations long before writing was invented. Just like a zip file on a computer, they kept the information organized, simplified, and sometimes hidden for protection.

We tend to view ancient peoples as inferior at distinguishing experience from imagination. Yet they lived closer to the sky than we do, both physically and psychologically. They tracked celestial bodies, aligned their cities with stars, and measured time by the movements of the planets. Their survival depended on paying attention to detail, pattern recognition, and memory.

Did they imagine gods when they saw beings coming down from the sky, bringing knowledge, changing bloodlines, making laws, or vanishing into fire? Or were they simply doing what humans have always done when they encounter something they don't completely understand: using the words they know best?

Religion has many contact stories. A series of meaningful encounters shaped by culture, fear, reverence, and power that collectively determine a profound influence on the people.

Gods weren't aliens, angels weren't spacecraft, and not every religious story is about a literal visit that needs decoding. Over-simplifying like that misses the point just as much as dismissing it outright.

Humanity has long engaged with something akin to consciousness, intelligence, or agency beyond the human realm. Our oldest myths are remnants of encounters, awaiting a culture capable of understanding them. Once you compare ancient traditions, the similarities become hard to overlook.

These cultures were divided by oceans, centuries, and different linguistic groups. They had no shared scripture, no global communication network, and no unified mythology. Yet they repeatedly described strikingly similar figures from above. Great beings associated with the sky, stars, or the heavens who interacted with humanity at key moments.

Sky Visitors

In Mesopotamia, the *Anunnaki* weren't abstract gods in the modern sense. They were depicted as powerful beings who descended from the heavens, governed the Earth, and imparted laws, technologies, and social order. Their stories are administrative and procedural, like official records wrapped in reverence.

In ancient Egypt, the gods were often depicted as emerging from the sky or dwelling among the stars. Pharaohs ruled and served as intermediaries, those who could communicate with the divine realm and, after death, ascend to join it. The *Duat*, the celestial afterlife, was carefully mapped, with gates, guardians, and pathways. This describes an ordered realm with rules and gateways.

Across the Vedic traditions of India, the Devas arrived in *vimanas,* flying vehicles capable of astonishing flight. People often see these accounts as just poetic symbolism, but their details are consistent enough to make you wonder: controlled flight, battles in the sky, weapons of light and sound, and folks watching from above. The language is mythic, but the perspective is technological.

In Mesoamerican civilizations, figures such as *Quetzalcoatl* and *Kukulkán* descended from the sky, brought knowledge, established laws, and promised to return. They were portrayed as teachers who walked among humans, intervened directly, and then ascended. Their arrival and departure were central narrative events in their culture.

Even in the Hebrew Bible, the word *mal'akh,* often translated as "angel", means *"messenger."* These beings arrive physically, deliver instructions, and depart. They eat, wrestle, instill fear, and produce tangible influence. Ezekiel's vision of the "wheel within a wheel," long considered purely allegorical, seems different when compared with other descriptions of luminous, moving structures in the sky.

What's striking is that belief in gods is universal across these cultures. It's interesting to see how their gods often behave less like all-powerful, all-knowing forces and more like local beings with their own personalities, coming and going, communicating selectively, and influencing human growth.

In these stories, the sky serves as both a stage and a place to start.

Mountains, ziggurats, pyramids, and temples were constructed to reach towards the sky, serving as connections between Earth and the heavens. Sacred sites were frequently aligned with celestial events, star systems, or planetary cycles, almost as if they were meant to serve as moments of contact. The divine manifested at particular locations, at designated times, under specific conditions.

This marks the transition of the concept of *sky beings* from conjecture to a more substantiated understanding.

Across cultures, these entities are associated with:

- descent from the heavens
- advanced knowledge and instruction
- authority regarding law, order, or civilization
- luminous or non-human appearance
- eventual withdrawal or disappearance
- promises of return

These motifs recur, encoded in myth and ritual.

Modern systems often use metaphors: the sky stands for transcendence; gods symbolize psychological archetypes; angels represent moral forces. While these ideas are helpful, they also suggest the experience is all in the mind and remove any sense of interaction or agency.

What if ancient people weren't inventing sky beings to explain natural features but were instead recording encounters with intelligences they couldn't understand through science?

If *panpsychism* holds that consciousness is fundamental

and widespread, engaging with non-human intelligence becomes not only possible but probable. In such a universe, the mind isn't confined to the brain, and intelligence doesn't have to be human-like to interact with it.

Humanity's first way to record contact with another *realm* was using story as a memory aid, ritual as a set of rules, and reverence as a means of management.

What were they responding to?

Our emphasis on metaphor has kept encounters camouflaged, and we are only now, technologically and philosophically, prepared to revisit them.

Altered States

Contact stories usually don't happen during normal waking moments. Whether called divine revelation, shamanic journey, prophetic vision, or strange encounters, they seem to depend on a change in how we see things.

Ancient religious systems clearly reflected this belief, in which interactions with gods, angels, or sky beings demanded special preparation. Purification rituals and intermediaries were involved in the intentional induction of altered states.

Fasting, chanting, sensory deprivation, isolation, and *psychoactive substances* were not just small rituals; they were ways to reach certain states. The experience could be risky, so rules and hierarchy were used to manage it.

These altered states served as gateways to various modes of perception. During these states, individuals reported experiences with autonomous intelligences, strange environments, and external information. The fact that these reports are consistent across different cultures suggests a repeatable phenomenology, despite differing interpretations.

Recent research has examined this pattern from a different

angle. Studies involving compounds such as DMT, *an endogenous molecule found in the human body,* have documented experiences in which individuals are immersed in coherent environments and interact with seemingly autonomous entities. Participants often report these encounters as more real than dreams, beyond their conscious control, and marked by an intense sense of something or someone.

Several experiments have begun examining longer altered states to see whether spending more time in them changes how interactive these perceived environments are. Early results suggest that the state's duration and intensity might influence how clear the experience feels. This is similar to older traditions that valued long rituals, initiations, and gradual immersion over quick intoxication.

Altered states are seen as doorways and thresholds for encounters. Ancient cultures understood these thresholds, while modern science studies them with neurochemistry and cognitive models. Both see the same boundary.

Sacred Fire

Across desert cultures, survival and revelation are often intertwined through the same substances. Food, medicine, and the sacrament were seldom seen as separate categories. When ancient texts describe divine nourishment falling from the sky or visions sparked by fire that does not burn, they may be describing experiences that are both physical and symbolic, felt in the body yet interpreted through cosmology.

One recurring hypothesis holds that the Bible's holy *manna* was not a miraculous confection in the modern sense but a naturally occurring organism, *possibly a desert fungus or lichen,* that appeared seasonally. These samples resisted easy storage and altered perception. Such organisms still exist in arid environments today, emerging briefly after specific climatic conditions and then vanishing.

To a nomadic people, its sudden appearance, nutritional value, and impermanence would have seemed otherworldly. The prohibition against hoarding may not have been moral at all but practical: the substance lost potency or spoiled quickly, reinforcing ritualized reliance on timing, obedience, and trust.

In the deserts of the Near East, certain resinous plants and psychoactive flora can induce intense visions when inhaled or ingested, often accompanied by a sensation of a radiant presence.

Forbidden fruit

Long before apples became the common symbol, the word "*fruit*" meant a product or yield. In many desert regions, thorny shrubs produce pod-like or cactus-like growths that contain alkaloids capable of inducing intense hallucinations.

Ethnobotanical records indicate that serpent-linked cults in ancient times venerated such plants not as recreational drugs but as portals. The snake, shedding its skin and moving between earth and shadow, became a symbol of knowledge gained through transformation.

In this interpretation, the *serpent* does not tempt humanity into moral failure; it introduces altered cognition. The knowledge gained concerns self-awareness, the sudden perception of nakedness, vulnerability, and separation. Once consciousness is altered, innocence becomes irrecoverable. The real punishment isn't banishment from a garden, but having to live forever in linear time.

Viewed this way, religion functions as a long-term contact protocol. Strict rules govern who may access altered states, when, and under what supervision. Prophets retreat to deserts or mountains. Initiates fast, isolate themselves, chant, or endure extremes before their revelation. The taboo is against uncontrolled access. When everyone can see ***behind the veil,*** authority disintegrates.

Earlier traditions that later seemed like metaphors may have actually been procedural knowledge, carefully recorded when sharing directly became risky. As authority centralized, spontaneous revelations became a threat. Mystical experiences were shared externally, delayed, or controlled by priests. Instead of divine presence showing through smoke or visions, it was communicated through texts, hierarchy, and beliefs.

Whether that *something* was internal neurochemistry, an emergent property of the brain, or an interface with an external intelligence remains an open question.

The rules emphasize how we access and view mythology, religion, and modern supernatural encounters as layered stories. These stories show how human consciousness has long interacted with things that go beyond ordinary perception. Altered states are like gateways, revealing patterns that people have described, ritualized, feared, and managed throughout recorded history.

Just as sacred substances were once reserved for initiates in rituals, today's most debated research areas, such as human consciousness, remote sight, and strange phenomena, are kept secret. The official reason is safety, but the real goal is control.

Declassified documents reveal that governments have repeatedly investigated consciousness as a strategic domain. Remote viewing programs, altered-state interrogation techniques, sensory-deprivation research, and psychoactive trials were pursued with the same seriousness as weapons development. We know consciousness can be altered, but can it be *controlled*?

Disclosure Dosage

Insufficient information breeds myths and conspiracies, while excessive information released without context can cause disaster. Ancient writings tackled this issue by conveying experiential knowledge through allegory.

Modern systems address this through redaction, gradual normalization, and selective leaks. The outcome remains the same: the population recognizes that something exists beyond the sanctioned narrative but lacks the means to engage with it directly.

Think about how we describe strange experiences today. People often talk about seeing strange lights, moving objects without engines, or feeling the presence of entities. These descriptions are similar to religious stories of visions, involving glowing lights, freezing sensations, and a powerful presence. Even though the words change, the basic events are the same.

Even modern disclosure follows a ritual design. It begins with preliminary acknowledgments, involves carefully selected intermediaries, oaths of secrecy, and consequences for transgression. The public isn't treated as participants but as an audience, gradually acclimating to a new reality. What was once considered initiation is now commissioning.

If consciousness serves as a point of contact, whether with non-human intelligence, a deeper reality, or the mind's emergent properties, unrestricted access can be dangerous because it is powerful and questions authority.

Chapter Eight

The Reset Hypothesis

I want to take a moment to speak directly to you, reader to reader. *What if our timeline is wrong?* How often do we rely on assumptions in the absence of evidence, and how quickly can those assumptions take over?

Younger Dryas

The *Younger Dryas* is one of those hypothesis-shattering revelations. It marks a sudden return to near-glacial conditions at the end of the last Ice Age, beginning about 12,900 years ago. Researchers were surprised by how quickly the climate changed. In geological terms, this was like slamming on the brakes.

Temperatures in the Northern Hemisphere plummeted over decades, perhaps even within years, disrupting a warming trend that should have continued smoothly.

The name itself derives from a flower: *Dryas octopetala*, a small, white, eight-petaled plant adapted to cold, tundra-like conditions. Its pollen appears abruptly and abundantly in sediment cores from this period, signalling a rapid return to harsh climates.

The world our ancestors were adapting to vanished, replaced almost overnight by one they weren't prepared for.

Dryas octopetala / White Mountain Avens

For any civilization, proto or otherwise, such a shock would have been existential. Ice sheets advanced, weather patterns destabilized, and sea levels, already rising from melting glaciers, surged in catastrophic pulses. Coastal homelands vanished, and survivors fled inland, carrying little more than their memories.

Global Flood Myths

Mesopotamia – The Epic of Gilgamesh

One of the oldest flood stories comes from Mesopotamia, found in the *Epic of Gilgamesh* and earlier Sumerian texts. In this story, the gods decide to destroy humanity because human activity has become too noisy and chaotic, reflecting an imbalance. The god *Ea* warns *Utnapishtim* to build a large boat, save his life, and survive a flood that will cover the world.

After the waters recede, survivors release birds to find land and offer sacrifices when the birds return. This story is much older than the commonly referred-to biblical account, showing that it did not borrow from scripture but comes from an earlier tradition.

Hebrew Tradition – Noah's Flood

In the Hebrew Bible, Noah's flood functions as a comprehensive reset, encompassing environmental, moral, and genealogical aspects. Humanity has become corrupt, violent, and irredeemable, prompting God to destroy nearly all life and restart with a righteous family. The ark acts as a vessel of preservation, safeguarding seeds, animals, and lineage through the catastrophe.

Following the flood, a covenant is established, signified by the rainbow, symbolizing both survival and a vow to avoid future destruction. It's often overlooked that this story appeared in a context where *Near Eastern* flood myths were widespread, suggesting that the Hebrew narrative is a theological adaptation of an ancient disaster myth, emphasizing law, order, and moral continuity after chaos.

India – Manu and the Fish

In ancient Indian tradition, the flood story concerns Manu, the ancestor of humanity. A small fish, later identified as an avatar of *Vishnu,* warns him that a big flood will wipe out the current age. Manu is told to build a boat to save life and sacred knowledge. When the flood comes, the fish helps guide the boat to safety, anchoring it to a mountain as the waters go down.

The story is cyclical: the flood ends one age and starts another. Civilization restarted, and knowledge is preserved through keeping it safe, a common theme in the Reset Hypothesis.

Greece – Deucalion and Pyrrha

In Greek mythology, *Zeus,* angered by human wickedness

and lack of piety, sends a great flood to cleanse the earth. Forewarned, *Deucalion* and *Pyrrha* survive by building a boat and floating until the waters subside.

After the deluge, they restore humanity not through ordinary reproduction but symbolically, by throwing stones behind them that transform into people. This suggests not literal rebirth but the reconstruction of society from the remnants. Like other flood stories, it shows survival, moral renewal, and the passing of culture through a sudden break.

Indigenous Traditions – Americas, Australia, Polynesia

Flood stories are found in many Indigenous cultures around the world. They often have similar features, but details vary by location. Many Native American stories tell of worlds destroyed by water, with survivors escaping to mountain jujm tops or floating refuges.

Australian Aboriginal stories recall ancestral lands submerged and law-keepers who preserve knowledge through songlines.

Polynesian myths recount vast oceanic expansions, lost homelands, and star-guided migrations following the rise of seas that swallowed earlier worlds.

These stories are often dismissed as symbolic, yet many originate in cultures with oral traditions dating back to the end of the Ice Age, when sea levels rose sharply. Their consistency suggests they encode lived experience, passed down because there was no other way to carry memory after a reset.

Flood

Separated by oceans and millennia, these cultures recall the same type of event: sudden inundation, near-total loss, and the preservation of knowledge by a small number of survivors.

History, as often taught, is viewed as a linear and progressive process. Humanity moves from ignorance toward enlightenment, dragging stone tools behind it until civilization finally emerges a few thousand years ago. This narrative provides comfort, but it is increasingly difficult to defend.

The Younger Dryas is one of the most violent climate upheavals in human history. Temperatures in the Northern Hemisphere plummeted within a few decades. Ice sheets expanded. Ecosystems collapsed as sea levels, which had been rising since the last Ice Age, surged even faster, erasing everything humanity had built before this period.

Meltwater Pulse 1B

Geological evidence indicates that sea levels didn't rise gradually; they rose abruptly. *Meltwater Pulse 1B*, approximately 11,500 years ago, may have caused global sea levels to rise by several metres in a very short period. Coastlines disappeared, and continental shelves were flooded. Entire regions capable of supporting dense human populations were submerged.

If an advanced civilization existed before this event, its cities would be hidden beneath the oceans, not in deserts or jungles. The hardest places on Earth for archaeological study are where Ice Age humans would have lived: fertile coastal plains, river deltas, and shallow shelves.

Atlantis

Many people see Plato's story of Atlantis as just a made-up story. But the way it's told is really interesting. Atlantis is described as a highly advanced seafaring society that was suddenly wiped out by floods and earthquakes. It's thought to have existed a long time ago, even before Plato's time.

Instead of imagining a single lost city, Atlantis could be seen as a symbol, a collection of faded memories from different sunken civilizations. Over thousands of years, oral stories about these places faded and became less clear, but the tales of disaster remained constant.

Flood myths are found worldwide, including in Mesopotamia, India, Mesoamerica, and Scandinavia. Cultures separated by oceans all remember the same event. *Something happened.*

Ancient Advanced Civilization

Advanced doesn't mean industrial, nor does it depend on steel, fossil fuels, or silicon. Instead, it focuses on organization, knowledge sharing, precise engineering, and long-term planning.

Evidence increasingly shows that humans during the Ice Age were not as cognitively primitive as once thought. Places like *Göbekli Tepe* reveal impressive structures built by societies labelled as hunter-gatherers, groups that, according to usual ideas, shouldn't have been capable of such teamwork or symbolic sophistication.

This creates a bit of a puzzle. Either our ancestors suddenly acquired advanced social and architectural skills, or those skills were already in place, and we're just seeing survivors piecing together parts of a forgotten world. The second option makes a lot of sense based on the facts.

The Reset Hypothesis

The Reset Hypothesis suggests that human civilization has experienced at least one major collapse so severe that later societies mistakenly believed their cultures were the beginning. Knowledge didn't come from nowhere.

After the *Younger Dryas*, humanity didn't start from zero but from a significant deficit. Survivors clung to myths, symbols, and partial techniques, lacking the infrastructure or population to sustain them. Over generations, memory evolved into religion. Engineering devolved into ritual, and technology became sacred but not quite reproducible.

Stone circles, pyramidal shapes, and astronomical alignments are found worldwide because they follow the same underlying blueprints, albeit imperfectly retained.

Convenient Fiction of "Primitive Ancestors"

Labelling ancient peoples as primitive serves a purpose. It preserves the illusion of continuous progress. It strengthens modern identity, and it sidesteps uncomfortable questions about what we lost.

If past civilizations achieved levels of sophistication similar to ours, collapse happens more often than you might think. It's a cycle that shows how vulnerable we are. The idea of a reset also challenges authority.

Labelling our ancestors as primitive is just a helpful story, a kind of blank slate that lets us see modern civilization as inevitable, unique, and falsely superior.

Current Relevance

The Younger Dryas serves as an example of rapid climate change. We've seen similar patterns before, such as sudden climate shifts, rising sea levels, and people moving to new areas. What makes our current time different is just how big these changes are.

If civilization didn't begin right after the Younger Dryas but instead continued in disorganized remnants, we might need to re-examine many of humanity's oldest stories. These stories might be memories passed down from a world we once shared, including tales of gods, floods, and beings descending from the sky.

Long before we knew about modern spacecraft, many ancient cultures told stories of glowing crafts, sky beings, and watchers who appeared during times of chaos. While these stories are often seen as symbolic, the similarities across cultures suggest they may share a common experience. Legends of flying discs, shining chariots, flaming shields, and floating lights appear in traditions that may even predate the climate disaster.

If such phenomena were observed by a pre-reset civilization, whether as natural events, advanced technologies, or non-human intelligence, later societies would have inherited only mythologized versions of them.

This new way of looking at the so-called "ancient astronaut" stories gives them a fresh perspective. After a disaster, people often misunderstood machines, intelligences, or strange phenomena as gods because they had lost the original context.

UAP mythology is humanity's longest-standing mystery. A

story of unresolved encounters passed down through civilization's amnesia.

Catastrophe

The most concerning aspect of the *Reset Hypothesis* is that civilization might never have fully recovered. Today's society sees itself as the pinnacle of human achievement, but many of its key traits are, in fact, reactive.

We tend to respond to crises with emergency measures, extract resources at unsustainable rates, and push technological progress without thinking about the wise use of those advancements.

In a world changed by disaster, our priorities must shift to focus on survival, control, and redundancy. As specialization becomes more important than working together, we see that knowledge isn't stored in our minds or communities but in external databases and institutions. Memories from personal experiences highlight the urgent need to regain their importance.

This shows what happened after the Younger Dryas. Large projects are gone, replaced by smaller, scattered efforts. Precision is still evident in specific areas, such as astronomy and stonework, but the interconnected system that linked them all is no longer there.

If humanity survives after a disaster, secrecy might become a permanent habit. When knowledge is scarce, and the memory of the collapse lingers, even without awareness, controlling information is very important. Institutions have been created to help prevent chaos.

Resistance to disclosure stems from fear of triggering another disaster and, deep down, a sense that it is already on the verge of collapse.

Mankind may not be a young species exploring the universe for the first time, but an ancient species awakening from a long period of *sleepwalking.*

Modern Psychology and NHI

If NHI has ever interacted with Earth, its relationship with us might be about keeping things in check. From a non-human perspective, especially one that considers long timescales, human civilization might look like a repeating cycle: growth, acceleration, instability, and collapse.

An NHI wouldn't need to destroy humanity to take control; it could just as easily interfere with it. A catastrophe achieves what direct control cannot: it damages memory while keeping the species alive. After a reset, humanity might forget what it was capable of, but still have the desire to rebuild. This leads to a civilization that advances technologically without fully reconnecting with its past.

This is how post-catastrophic psychology explains what happens. Trauma makes it hard to focus, so survivors tend to pay attention to authority, rituals, or explanations rather than trying new things.

In these situations, strange experiences are often dismissed as myths or attributed to fear. Non-human presence might be thought of as divine, evil, or just unreal. People may sense these experiences, but they're usually hard to understand or make sense of.

From an NHI perspective, this brings several nice advantages. Humanity stays ecologically contained, limited by forgotten technological thresholds, and psychologically depends on mediated truths. Contact can be somewhat ambiguous, so uncertainty can serve as a way to manage it.

Who stands to gain the most from this arrangement? NHI

benefits by ensuring the long-term stability of the system it inhabits or monitors. But human institutions also reap rewards, especially those that emerge immediately after a collapse.

Priesthoods, monarchies, and later technocracies flourish in post-reset societies because they inherit narrative authority. They serve as intermediaries between the unknown and the population, echoing an age-old pattern: access must be earned, filtered, and tightly controlled.

This describes a management system with multiple levels. NHI sets the maximum limit for human growth. Human authorities control access within that limit. Neither uses force because memory loss helps them.

The biggest surprise isn't that NHI interferes, but that it always has. What we sometimes call gods, watchers, or anomalies aren't just visitors; they're constant presences, working through resets and changing their ways as humanity forgets and relearns.

Throughout history, whenever humanity makes progress in technology or psychology, the system seems to forget.

Chapter Nine

Celestial Megaliths

I travelled to Scotland in June 1996.

There wasn't a grand plan for the trip, no tour group or list of must-see sights. I needed to think about my situation and where my life was headed, so I hastily decided to go somewhere. Scotland was at the top of my list. I called my dad, as I often did, to share the news. He responded right away, his voice serious. He told me, "Go on the trip, Jody, don't just talk about it."

This was my very first trip abroad! With just a bit of savings, I flew with British Airways from Toronto, making connections through New York and Birmingham before finally arriving in Glasgow. From there, I hopped on a train heading north to the charming Highland town of Kingussie, where I stayed at a cozy bed-and-breakfast run by Joan, an Englishwoman who had retired to Scotland and found peace among the hills.

She showed me warmth and patience. Joan had the time, curiosity, and willingness to take a solo traveller along narrow Highland roads, gently swerving and occasionally cursing when a rabbit darted in front. She called them a menace, shaking her head in exasperation, as if they were just a nuisance in the landscape.

Of course, I made sure to visit the castles! I enjoyed a Jacobite cruise in Inverness and checked out the Loch Ness Monster exhibits just like everyone else. I loved soaking up the rich history Scotland proudly shares. On top of all that, there was also a peaceful, quieter side that called to me.

My curiosity led me here because of the *stones.*

Years earlier, I had received a book as a high school graduation gift: *Outlander* by Diana Gabaldon. The novel portrayed stone circles not merely as scenery but as thresholds, places where time warped, memories lingered, and something older than history still waited. I explained this to Joan; she understood, then said she knew a place.

Like all lasting myths, stone circles hinted at a deeper meaning no longer fully understood. They weren't marked on the usual tourist maps, nor did they have souvenir shops or guided reenactments. It was a calm, peaceful site near Inverness called the *Clava Cairns.*

Cairns of Clava

When we arrived, the landscape immediately felt different. It wasn't anything too grand or overwhelming, but you could tell it was definitely intentional. Low stone rings sat comfortably into the earth, their walls curving inward like cozy igloos made from heavy rock. These weren't like Stonehenge, massive monuments meant to dominate the horizon. Instead, they were arranged thoughtfully and gave off a very warm, intimate feeling.

The stones were ancient, much older than the medieval castles Scotland is known for. These cairns were not haphazard piles of stones. These formations were meticulously designed, systematically positioned, and intentionally placed. They reflect a purpose rooted in complex astronomical, mathematical, and environmental knowledge that exceeds our typical understanding of early human capabilities.

Standing there, I felt something I couldn't quite describe at the time: that these structures weren't the beginnings of simple architecture but leftovers from a system already fading away.

Only years later did I understand a deeper layer of what had happened and its effect on me.

The stone circle in *Outlander*—Craigh na Dun—is fictional. However, it was inspired by real sites like this one. When the television series based on the book series aired decades after my visit, Clava Cairns became a primary real-world stand-in for that myth. Thousands of people later came here seeking what the story promised: a thinning of time, a doorway, a crossing moment.

In 1996, none of that was around yet. There were no crowds. I arrived alone, driven by curiosity and a sense that these stones held importance I didn't fully grasp.

That realization still gives me chills. I didn't visit Clava Cairns because it was famous for a time-travel story. It became famous for that story *after* I had already been there. That raises a difficult question: *do stories turn places into thresholds, or do they gather around places that already serve that purpose?*

This chapter explores sites such as Clava Cairns, as well as even older, larger, and more well-defined ones. From the stone circles across the British Isles to *Göbekli Tepe*, and from pyramids aligned with celestial cycles to megaliths built with tolerances that modern engineers still find fascinating, we're often asking the same question:

How did ancient people acquire so much knowledge, and why does that knowledge seem to appear suddenly, only to fade away?

To answer that, we need to look beyond seeing stones as monuments. Instead, let's see them as tools, as technology, and as symbols of a deeper, lost connection between humans, the sky, and the passage of time.

Oldest Known Site

Located in southeastern Turkey, *Göbekli Tepe* predates the renowned *Stonehenge* by over six thousand years and even the Great Pyramid by a greater margin. Its construction, dating back to a time when conventional history suggests humans were still nomadic hunter-gatherers, lacking agriculture, metallurgy, writing, or complex social structures, challenges long-held assumptions about the development of human civilization.

Monumental Archaeological Site

Large T-shaped pillars, some weighing many tonnes, were carefully quarried, transported, carved, and erected with precision. They are arranged in circular enclosures, their surfaces decorated with animals, symbols, and abstract designs whose meanings are still debated. The pillars are smooth and intentional, showcasing a uniform and consistent appearance.

The sequence of events is quite fascinating. It appears that Göbekli Tepe was constructed before the advent of agriculture, not after. There's no sign of permanent homes, farms, or towns from the early times that would need a lot of labour. Instead, it seems that people mainly gathered there for rituals and ceremonies, to connect with something beyond everyday survival. Only later did settled communities develop.

This challenges the usual idea. Usually, we think humans first settled, farmed, built cities, and only later developed religion, astronomy, and big monuments. But Göbekli Tepe suggests the opposite: shared symbols, teamwork, and understanding of the cosmos might have been what first brought humans together in large groups.

Like Clava Cairns, Göbekli Tepe is neither a fortress nor a palace. It's a place of gathering, enclosure, and orientation, a site designed to highlight the connection among people, animals, seasons, and the sky.

And then, just as intentionally as it was constructed, it was buried.

The enclosures were filled with rubble and earth and carefully preserved, as if the builders knew the site might be needed again. This was a deliberate act, offering a sense of closure, protection, or transition.

When seen alongside later megaliths from the British Isles to the Nile, Göbekli Tepe felt like the beginning of a longstanding tradition, one in which stone was used for communication.

Sky Becomes the Blueprint

Megalithic monuments take on a special meaning when

you think about the sky. Across different continents and throughout history, these sites share a common feature. Passages, pillars, corridors, and sightlines are purposefully aligned with celestial events like solstices, equinoxes, lunar standstills, and star risings. They're built with careful attention to direction, aiming towards the sky.

At *Clava Cairns* in Scotland, the entrance passages and graded standing stones frame the midwinter sunset, the darkest moment of the year, when the sun seems to die before returning.

The Salisbury Plain in England is home to *Stonehenge*, whose main axis aligns with the summer solstice sunrise and the winter solstice sunset. *Newgrange*, older than Stonehenge, channels the winter solstice sunrise down a narrow passage so precisely that sunlight reaches the inner chamber for only a few minutes each year.

Göbekli Tepe in southeastern Turkey pushes this pattern further back in history. While debate continues over which specific celestial targets were intended, several leading hypotheses converge on the same conclusion: the builders were tracking meaningful, repeatable patterns in the sky.

If some enclosures align with *Sirius*, this would place their design in the early Holocene, roughly 9300–9000 BCE, when the star reappeared after centuries below the horizon—a dramatic and rare event that could have marked seasonal or ceremonial timing.

Others point to alignments with *Orion* and related stellar groupings, whose southeastern rising at *Göbekli Tepe* would have occurred between approximately 9300 and 8800 BCE. These stars, associated with hunting, death, and regeneration, would have been visible throughout the winter night sky, offering repeatable celestial markers across generations.

Equally interesting are interpretations that focus on the sun and moon. The circular enclosures might have served as horizon calendars, helping to mark solstices, equinoxes, or lunar standstills by tracking sightlines over time.

To accurately track solstices, observations of the sun's extremes need to be made over many years. Predicting lunar standstills requires documenting an 18.6-year cycle.

Recognizing precession, the slow wobble of Earth's axis that gradually shifts the background stars, requires knowledge maintained for periods much longer than a single lifetime.

Precession

This sky-ground conversation goes beyond Europe and the Near East. At Chichén Itzá in the Yucatán, the pyramid called *El Castillo* is built so that during the equinoxes, light and shadow work together to make it look as if a serpent is sliding down its stairs. It's a special event linked to the calendar, celebrating planting seasons and spiritual renewal.

* * *

In China, ancient ceremonial centers like the *Temple of Heaven* in Beijing were meticulously aligned with the cardinal directions and solstitial cycles, embodying a cosmology where earthly order was fundamentally linked to harmony with the heavens. Earlier sites and imperial capitals were deliberately arranged according to celestial grids, integrating astronomy deeply into governance, ritual, and spatial organization.

* * *

Across India, the Jantar Mantar complexes are charming stone observatories that turned architecture into astronomical tools.

Built to track solstices, equinoxes, eclipses, and planetary movements, they demonstrate that long-term skywatching can be impressive.

* * *

In parts of Russia and Central Asia, there are some clear examples. For instance, the *Arkaim settlement in the southern Urals* (around 2000 BCE) has circular walls aligned with the solstices, suggesting ritual and agricultural significance.

The megaliths and petroglyphs in the Bishkek Valley and the Altai Mountains feature carvings marking sunrise and sunset points, as well as lunar standstills, linking human activities to regularly recurring celestial events.

* * *

In North America, including Canada, Indigenous stone features such as medicine wheels, alignments, and ceremonial landscapes were thoughtfully aligned with solstices, star risings, and the cardinal directions, knowledge passed down through oral tradition.

Materials, languages, and myths differ, but the main goal is the same: to link human life to the sky's steady motions and to create lasting symbols of this bond.

Pyramids and Orion

No ancient structures have been examined, measured, debated, and mythologized more than the pyramids of Giza. For centuries, they have fascinated humanity with their size and with what appears to be embedded within them: an objective that goes beyond Earth.

The Great Pyramid of Khufu has narrow internal shafts extending from the King's and Queen's Chambers, features that are often mistaken for ventilation shafts. Their true purpose became clearer only after their alignments were carefully measured. These stellar shafts are aligned with stars that were significant at the time of construction but are no longer so.

The southern shaft of the King's Chamber aligns closely with *Alnitak*, one of the stars in Orion's Belt, while the northern shaft points toward *Thuban*, which was the pole star at that time. Today, Thuban no longer marks true north, and Orion's Belt appears in a different position in the sky relative to Earth.

This shift happens because of *axial precession,* a slow wobble of Earth's rotational axis that occurs over approximately 26,000 years. The key implication is that the builders aligned the pyramid not to a fixed point in the sky, but to the way the sky appeared in their era, demonstrating a deliberate and insightful understanding of celestial movement.

Beyond the internal shafts, the layout of the three main pyramids has drawn significant attention. When viewed from above, the pyramids of Khufu, Khafre, and Menkaure closely resemble the three stars of Orion's Belt in their relative positions and spacing. This idea is often called the *Orion Correlation Theory.*

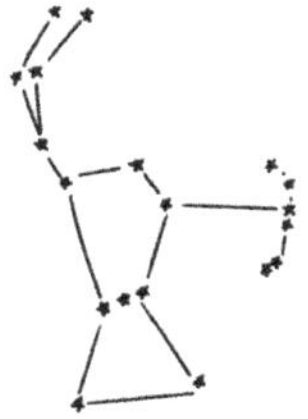

Orion Constellation

Orion was not merely a random constellation. In ancient Egyptian cosmology, it was associated with *Osiris*, the god of resurrection and the afterlife. The pharaoh was believed to ascend to the afterlife and join Osiris among the imperishable stars. Aligning a royal tomb with Orion served as a navigational aid.

The Nile was often likened to the Milky Way, which was envisioned as a celestial river through which souls journeyed. In this view, the Giza plateau becomes a terrestrial mirror of a cosmic map, linking earth, sky, death, and rebirth into a unified system.

Whether one regards the Orion correlation as an exact design or as an emergent pattern supported by myth, the core fact remains: the pyramids are linked to the sky. Their primary alignment is remarkably precise, deviating from true north by only a fraction of a degree, an accuracy comparable to modern surveying techniques.

Pyramids

Such exactness demands long-term observation, consistent reference points, and methods capable of translating astronomical knowledge into monumental architecture. Like memory devices built to last through dynasties and messages carved in stone for a future that might forget how to read the sky.

Megaliths aren't so much temples as helpful tools that guide us in understanding how the cosmos moves. They help communities connect with larger cycles that extend beyond just a single lifetime.

What's really impressive is this level of precision. Even long before the days of writing, metal tools, and formal math, humans seemed to intuitively understand how simple shapes connect to the movements of the universe.

Temples in Egypt and Mesoamerica were periodically rebuilt or reoriented, as if to keep pace with a sky that was slowly drifting. This suggests a dedication to staying aligned with changes in axial precession.

Stonehenge

If the pyramids speak the language of stars and eternity, *Stonehenge* speaks in cycles.

Rising from Salisbury Plain, Stonehenge has fascinated people for centuries, and it still puzzles experts today. It wasn't built all at once but in stages over more than a thousand years. This shows that Stonehenge was a *project* that many generations maintained, modified, and reinterpreted over time.

Stonehenge is a device for marking time.

Stonehenge

Its most well-known alignment is with the summer solstice sunrise, when the sun rises directly over the Heel Stone, lighting the centre of the circle. Less celebrated but arguably more significant is its alignment with the winter solstice sunset, a moment linked to death, dormancy, and eventual renewal. Evidence suggests that this darker turning point may have been the primary focus of ritual activity. Like *Clava Cairns* and *Newgrange*, *Stonehenge* emphasizes the moment when light fades and returns.

Massive sarsen blocks were transported over twenty miles, while the smaller bluestones originated from the Preseli Hills of Wales, almost 150 miles away. Moving them required careful coordination, planning, and perseverance through landscapes that didn't offer clear practical benefits for the effort.

What often goes unmentioned is the site's geometric sophistication. Stonehenge incorporates circles, horseshoes, and axial symmetry. The lintels are curved to follow the circle's arc and are locked into place with mortise-and-tenon joints, techniques more commonly associated with advanced carpentry than with prehistoric stonework.

Stonehenge also encodes longer cycles. Some researchers propose that the layout of stones and holes could be used to track lunar eclipses and the 18.6-year lunar standstill cycle. Regardless of whether all proposed functions are accepted, the main pattern is evident: this site was built to observe recurring celestial rhythms over long periods.

Unlike the pyramids, Stonehenge is open to the sky and accessible to everyone who visits. It isn't a sealed chamber or a royal tomb but a communal, exposed space that invites participation. Any knowledge associated with it was intended to be shared, reenacted, and remembered by the community.

Life unfolds in large patterns that extend beyond individuals, and survival depends on understanding your place within them.

Part Four

DARK SIGHT

Chapter Ten
Invisible Frontier

Most popular stories begin in the United States, not this one. It starts with the Soviet Union. In the 1960s and 1970s, U.S. intelligence analysts discovered that the Soviets were conducting extensive research into telepathy, psychokinesis, and bioenergetic fields. Laboratories across the Eastern Bloc were exploring consciousness.

The concern in Washington was whether the Soviets believed these abilities were real, and whether they were using them to spy. In intelligence work, belief alone can be operational. If your opponent believes a tool works, you can't afford to ignore it.

Remote Viewing

Remote viewing is the ability to sense distant places, objects, or events through imagination rather than the physical senses. Many cultures have described similar practices. In various shamanic traditions, from the Himalayas and Siberia to

Africa and India, spiritual healers said they could enter altered states to observe or gather information about distant locations.

In early modern Europe, some thinkers, such as *Emanuel Swedenborg*, wrote about vivid impressions of distant events that were experienced in trance-like states. Later, during the rise of mesmerism in the 1700s and 1800s, some hypnotized people described hidden or faraway scenes in great detail. By the late 1800s, members of the *Society for Psychical Research* in London were experimenting with what they called "travelling clairvoyance," in which participants, with their eyes closed or covered, tried to describe unseen targets.

Stargate

That mindset sparked what would later be known as the *Stargate Project*: a classified, decades-long investigation into a phenomenon described in intelligence documents as remote viewing. The supposed ability to perceive distant or hidden information using only the mind.

To modern readers, the idea that U.S. intelligence agencies once funded "psychic spies" can sound like Cold War fantasy. But *Stargate* was real. It is documented in declassified CIA files, Senate reports, and Defence Intelligence Agency memoranda. It ran for more than twenty years, survived multiple reorganizations, spanned military and civilian intelligence channels, and persisted well beyond the height of the Cold War.

Can human consciousness access information beyond the limits of space and time? According to those who ran the program, the answer was complicated.

Spy

Awareness

The earliest formal research was conducted at the *Stanford Research Institute* (*SRI*) under the supervision of physicists *Russell Targ* and *Harold Puthoff*. Their team brought together an unusual mix of participants, including scientifically trained observers, experienced meditators, individuals reporting anomalous perceptions, and classified military personnel.

Under controlled conditions, subjects were asked to describe distant locations, hidden facilities, or sealed targets. Early sessions startled even seasoned analysts. Some participants sketched diagrams later confirmed by satellite imagery. Others described geographical features, layouts, or activities thousands of kilometres away with alarming accuracy. The CIA continued to fund the work.

What followed was a long line of classified projects, each building on the last and moving between agencies under rotating code names: *Scanate, Gondola Wish, Grill Flame, Center Lane, Sun Streak,* and finally *Stargate*. The names might have changed, but the projects were the same.

One of the most consistent findings across the programme was something deceptively austere: *arousal state*. In psychology and neuroscience, arousal is the level of nervous system activation. High arousal elicits stress, increases adrenaline, promotes rapid thinking, and triggers the fight-or-flight response. Medium arousal corresponds to normal waking alertness. Low arousal, by contrast, denotes calm, inward-focused, meditative states.

Stargate researchers such as *Puthoff, Targ, Edwin May,* and *Ingo Swann* demonstrated that remote viewing accuracy can be significantly enhanced when participants achieve *low-arousal states.* Conversely, high arousal, which activates logic and problem-solving, distracts from the subtle impressions essential to successful remote viewing sessions.

When participants remained calm, sensory distractions decreased, heart and respiratory rates slowed, and brain activity shifted to the alpha and theta bands. Once the brain waves entered these bands, symbolic and visual impressions emerged inherently. In short, the mind performs best when it stops trying.

Over time, consistent patterns resulted across CIA, DIA, and Army Intelligence assessments. Remote viewing sessions were occasionally used to identify hostages, describe foreign installations, or locate missing assets. Accuracy varied: some sessions failed entirely, while others were remarkably precise.

Distance didn't reduce effectiveness; targets on different continents or in different time periods showed results similar to those of nearby ones. Quieting the mind and meditation consistently boosted performance. However, no known physical explanation could account for this. Still, the operational data was convincing enough to be used in real missions.

Some analysts viewed *Stargate* as a revolutionary tool that required no satellites, intercepts, or boots on the ground. Others considered it unreliable or politically dangerous. Yet one fact is hard to ignore: *no intelligence program survives for more than twenty years without producing some useful outcomes.*

Despite funding issues, internal doubts, and external criticism, *Stargate* kept going until 1995, well after the Cold War ended. When it was declassified, officials said remote viewing showed some statistically significant results but wasn't reliable enough for ongoing use.

Unofficially, many involved disagreed. Nearly every successful remote viewer reported similar internal experiences: deep calm, symbolic imagery, hypnagogic visuals, a sense of bodily detachment, sudden clarity, or fully formed impressions. These experiences closely paralleled shamanic journeying, psychedelic states, lucid dreaming, near-death experiences, and mystical visions described in religious traditions.

After 1995, much of the research moved to private companies such as *SAIC and SRI International,* aerospace contractors, and DOD-connected think tanks. They continued to study consciousness, clairvoyance, sensory deprivation, and human behaviour, often under different names. In the late 1990s and 2000s, new programs emerged, including *Human Systems Integration, Cognitive Engineering, Enhanced Situational Awareness,* and *Non-Kinetic Intelligence Platforms.*

DARPA

DARPA (*Defense Advanced Research Projects Agency*), originally founded as *ARPA* in 1958, naturally became the hub for high-risk research into cognition and perception. Although it does not claim to have run remote viewing programs, it funds research on *brain-computer interfaces, cognitive state modulation, intuition enhancement, subconscious threat detection,* and *predictive perception analytics.*

Many researchers describe these as Stargate reinterpreted through a neuroscience lens. The term "anomalous cognition" continues to appear in contractor white papers, FOIA-released performance studies, intelligence lexicons, and private-sector training programs for DOD clients.

Former participants, including *Ingo Swann, Joseph McMoneagle, Ed Dames,* and *Paul Smith*, have publicly stated: "The unit ended, but the research didn't." They believe it moved into deeper departments, such as special-access programs, private labs, and contractor-run initiatives that operate beyond congressional oversight.

Whether Stargate continues in any official capacity remains unconfirmed. Nonetheless, its documented legacy is undeniable. It stands as a rare, verified example of government bodies venturing into areas usually associated with mystics, shamans, and theologians.

Plausible Deniability

Over time, research in perception areas started to move towards contractor-run facilities. These labs needed to be flexible, fast and discreet qualities that can be hard to achieve in typical government or academic settings. Private labs could bring together experts from different fields, test new ideas, and adapt quickly, all while working on classified projects.

Contractors also offered a sense of *plausible deniability*. Programs could be spread across different labs, making it harder to see the full picture while still preserving the knowledge. Former participants call this a "fractured relay," in which projects that were once centralized within programs like Stargate now continue as separate, private experiments.

Research into extraordinary perception and predictive thinking isn't subject to oversight. In the contractor ecosystem, work can proceed on its own terms, judged by usefulness and not public regulations.

Contemporary Edition

By the early 2000s, remote viewing had largely faded from public intelligence discussions. The intelligence community uncovered a crucial insight: perception-based tools couldn't be manipulated as hardware could. They demand the right mental states, and most importantly, the absence of pressure to perform.

Today, no known government openly admits to operating a remote viewing unit. Yet, around the world, research into non-local perception, intuition, predictive cognition, and anomalous awareness persists. What was once called "remote viewing" is now referred to as "anomalous cognition," and "psychic functioning" has been rephrased as "pre-conscious signal detection." The subject has been reclassified linguistically, making these phenomena more credible.

Civilian organizations now train individuals using protocols directly derived from declassified military procedures. These groups emphasize disciplined sessions in blind targeting with strict feedback controls.

Internationally, research into extraordinary perception continues. In the United Kingdom, academic societies continue to study non-local consciousness within experimental frameworks that are over a century old.

In certain regions of Europe and Asia, consciousness research institutes examine intuition, meditative perception, and altered cognitive states as performance factors rather than as paranormal curiosities.

Russia and China, although largely opaque, still publish occasional academic material on human biofields, intuition, and nonlinear cognition. These publications are within neuroscience or systems theory, obscuring any link to intelligence gathering.

Recent research supports Stargate's final results: accuracy exceeds chance, with some individual differences, and performance depends more on mental state than on effort or intelligence. Remote perception seems trainable to some extent, but it isn't consistent. Some people show repeatable success, while others don't, regardless of instruction.

In civilian settings, remote viewing has been used experimentally for archaeological discoveries, missing-person cases, financial predictions, and environmental studies. Reports are cautiously qualified because no reputable organization claims remote viewing is infallible. Instead, it is viewed as a supplementary tool that refines search areas and is used with traditional methods.

Modern programmes recognize something early Stargate struggled to formalize: the observer effect. The act of observing seems to influence the system being observed. Targets that are emotionally charged, strategically important, or already under intense scrutiny tend to leave stronger impressions. Neutral targets don't.

Remote viewing suggests that consciousness receives and interacts with informational environments. If so, observation isn't passive; it's relational. This insight suggests a spooky parallel.

Observation

Unidentified aerial phenomena typically appear silently and vanish without a trace. These objects don't behave like typical reconnaissance aircraft that systematically map terrain or linger for detailed measurements.

This behaviour is similar to intelligence gathering, but not in the usual manner. While Stargate seemed like an attempt to observe from afar, UAP behaviour appears to do the opposite: it approaches, and simply paying attention seems to trigger it.

Nuclear facilities, military drills, and periods of heightened geopolitical tension often attract attention. UAP encounters suggest foreign surveillance. The intelligence community faces a dilemma: how should it respond to an observer it cannot deter, communicate with, or reliably measure?

So far, the answer has been silence because acknowledging it would require redefining perception itself as a domain of engagement.

Civilian

Today, anyone can learn remote viewing directly through protocols based on declassified intelligence manuals. Training emphasizes mental discipline, low-arousal states, organized data collection, and delayed responses. Some practitioners achieve consistent results, though many do not. The variability remains an unresolved issue.

Now that remote viewing is unclassified and openly accessible, its impact on security remains uncertain. It's clear that the invisible frontier Stargate crossed was the realization that perception might not be limited to the physical body.

Weak Channel

As remote sensing advanced, researchers began to notice unexplained results. The most precise sessions didn't seem to be conscious perception. Instead, they resembled memories. Viewers often described impressions arriving suddenly and effortlessly, then vanishing as soon as they tried to examine them too closely.

This created an uncomfortable overlap with an older human faculty: *dreaming.*

Dreaming

During sleep, the mind often produces imagery that isn't constrained by time, place, or logic. Dreams flow seamlessly through memory, imagination, and emotion, often blending past and future into a single symbolic narrative. Occasionally, these stories seem to predict events that haven't yet occurred.

Cultures throughout history have recorded such experiences under various names, including omens, visions, warnings, and visitations. The phenomenon remains remarkably consistent.

Researchers have acknowledged this similarity. The mental states most conducive to effective remote viewing closely resembled those of dreaming. Brainwave patterns aligned, and symbolic distortion was frequent. Direct interpretation often proved inadequate, whereas metaphors conveyed meaning more effectively than literal descriptions.

Premonitions follow a similar pattern. When people report knowing something before it happens, the information rarely materializes as a clear statement. Instead, it appears as a feeling, an image, a dream, or an inexplicable sense of urgency. The conscious mind struggles because the signal arrives in the form of symbols.

This might explain why intelligence agencies found it tricky to put remote viewing into practice. The mind seems capable of accessing information beyond what our common senses pick up, but it does so more for survival and understanding. Dreams and premonitions tend to act as warnings rather than offer clear explanations.

Stargate asked the dreaming mind to act like a machine. The intelligence community eventually moved away from this approach because consciousness revealed too much, without providing any logical insight. Dreams can't be commanded, nor can premonitions be ordered.

The channel endures, as each night, the human mind rehearses futures, revisits the past, and explores realities beyond physical limits. Occasionally, synchrony occurs, and the dream reflects an event. Other times, a feeling signals an impending crisis because somehow awareness surfaces.

It's unclear whether this is just a coincidence, a subconscious guess, or a deeper non-local process. But the fact that the experience kept happening suggests that whatever Stargate involved was not limited to trained viewers or secret programs.

Consciousness can perceive beyond the present moment under the right conditions, when we stop forcing it.

Patterns

There are no confirmed cases of communicating with NHI through remote viewing programs. After many years of testing, intelligence agencies have never documented a session with clear conversation, useful knowledge, or reliable interactions.

However, some viewers experienced brief, symbolic impressions of non-human presences and environments unlike any known terrestrial setting. Others described feeling observed in return. These fleeting sensations were subtle, mysterious, and impossible to verify, so they were not acknowledged formally.

From an intelligence perspective, this ambiguity was quite problematic. When information can't be clearly distinguished, it becomes difficult to take action. The programs weren't intended to delve into metaphysics; their purpose was to deliver clear answers.

In many incidents, people sense an unwelcome presence. If non-human intelligence is real, it might not communicate with language the way humans do.

Remote viewing might have limits that allow observation but block communication.

Chapter Eleven

Psychedelics and Contact

Psychedelic compounds cause hallucinations and disrupt how we filter reality. Neuroscience recognizes that the brain naturally suppresses sensory and cognitive information to help us stay focused. Psychedelics reduce that suppression, making more information come through.

Across different cultures, users describe experiences such as out-of-body experiences, distorted perceptions of time, perceived interactions with autonomous intelligences, and environments inhabited by strange beings.

From an intelligence standpoint, this is significant. These experiences are similar to reports from remote viewers in deep, low-arousal states and from individuals experiencing spontaneous telepathy.

When a mind is freed from its usual filters, it seems to *absorb* more incoming data.

Dream Interface

Dreaming represents humanity's oldest altered state, and possibly its most underappreciated.

Unlike waking imagination, dreams often let go of our usual sense of control. The analytical mind steps back, sensory input softens, and logical thinking slows. What remains is a rich, immersive experience that feels completely unbound by physical limits.

Nearly every culture records *dream visitation*:

- Ancestors delivering warnings
- Spirits imparting instructions
- Entities conveying future events
- Knowledge mysteriously acquired

In many traditions, dreams weren't seen as symbols. They were believed to be literal encounters with another realm governed by different rules. Modern psychology views dreams as internal processes, and within scientific research, premonitory dreams remain difficult to ignore.

Intelligence agencies discreetly monitored these premonitions, recognizing that dreams can occasionally provide valuable information. Similar to remote viewing, dreaming tends to be most effective when the sense of self is suppressed.

Question of Source

Where does all that information come from? These experiences have key features, including acting on their own, clear communication, sharing knowledge they don't personally possess, and a sense of being controlled by an outside force.

Whether perceived as spirits, non-human intelligences, subconscious pattern recognition, or something else entirely, the phenomenology remains consistent. The experiencer doesn't feel like an author; they feel like a participant, and this distinction is important.

If consciousness can access information without relying on sensory input, and if that access becomes easier when the sense of individual identity lessens, then the mind might act more like a receiver than a generator. This idea supports both ancient rituals and modern intelligence studies.

In the twentieth century, intelligence agencies rediscovered techniques like remote viewing, dream-based intuition, shamanic journeying, and psychedelic experiences.

Ancient practitioners saw their experiences as a relationship with otherworldly forces, while modern researchers treat them as bizarre data to be collected.

Altered States

Psychedelic compounds hold a special place in this conversation because they affect the way our consciousness filters information. Substances like psilocybin, DMT, and LSD tend to decrease activity in the brain's default mode network, which is the part responsible for keeping our sense of self, our story, and our thinking in check.

When this network is supressed, people often feel a loss of boundaries around themselves, along with a sense of greater meaning, better pattern recognition, and sometimes experiences with strange beings.

Communication seems clear because information is shared symbolically. Often, people share insights or knowledge they weren't consciously aware of, perceiving these as coming from outside sources rather than from their own imagination. It seems that the brain is busy generating, decoding, or making sense of its content.

This has led some researchers to revisit the pineal gland,

which *Descartes* once called the "*seat of the soul*" and which was regarded as important in many ancient spiritual traditions. The pineal gland helps control our sleep-wake cycle and produces melatonin. It is also a key part of the brain's network, with many connections.

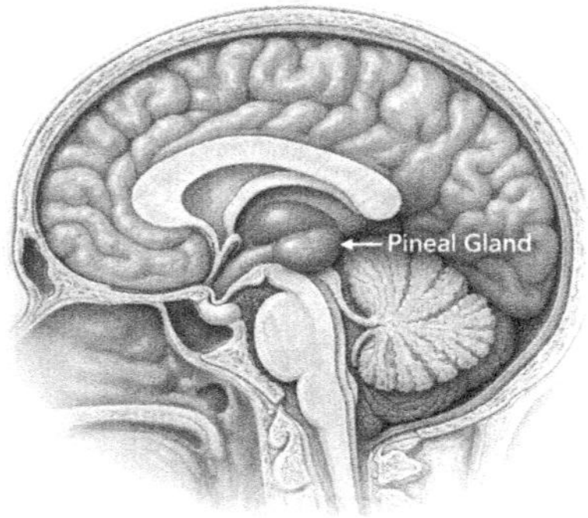

Modern ideas suggest that the pineal gland might produce natural psychedelics, especially DMT, which is found in tiny amounts in mammalian brains. Whether the pineal gland acts as a literal transmitter or not, its symbolic meaning reflects a shared belief across many cultures: that different states of consciousness can grant us access to or awareness of certain experiences.

Ritual darkness, fasting, rhythmic stimulation, dream incubation, and psychoactive plants all produce the same biological effect: suppression of ego dominance and filtering sensory input.

Psychedelic states involve temporarily shutting down the parts of the brain that filter perception. When the mind releases its ego, it's capable of perceiving information that seems external, independent, and meaningful.

Ancient cultures saw these experiences as real relationships with spirits, ancestors, or gods. Today, many see them as internal hallucinations. Neither viewpoint explains how consistent, detailed, and repeatable these events are across different people, cultures, and times.

Contact

In modern UAP reports, a common yet little-studied pattern is that altered states of consciousness often precede contact. Witnesses often record encounters occurring during sleep, deep meditation, emotional shock, illness, sensory deprivation, or right after using psychedelics. These are times when normal perception is weakened.

Timing really matters. Usually, contact doesn't happen when someone is highly alert, stressed, or deeply focused. The nervous system is closed and unreceptive during these times.

These states involve less activity in the brain's default network, changes in how the thalamus filters signals, and allow the circadian chemistry regulated by the pineal gland. Perception becomes less connected to external senses and more influenced by internal signals.

UAP behaviour follows this selective accessibility. Encounters are usually brief, personal, and subjective, even when physical signs are seen. Communication is often symbolic, emotional, or telepathic. Witnesses often say they feel watched.

This reflects how psychedelic and dream encounters often seem clear, purposeful, and coming from outside sources, no matter what they truly are.

Observers

This might explain why traditional detection and contact methods often don't work. If interaction relies more on perception than on physical closeness, then radar, visual monitoring, and intense pursuit are not practical tools. The signal might not be coming through the senses at all.

Ancient cultures understood contact as a relationship influenced by gateways and altered states. Modern science, based on materialist ideas, often dismisses these experiences as hallucinations.

If non-human intelligence communicates through consciousness, then altered states of mind become a means of accessing information. The pineal gland, our internal clock, and the suppression of personal identity may serve as biological tools to temporarily facilitate perception and the connection of information.

In this case, UAP encounters break a boundary that normally helps preserve our mental stability.

Constraint Hypothesis

Why Sustained Contact Is Avoided

One reason encounters with NHIs are brief and unpredictable is the human mind's limits. Interaction could be very risky.

Human consciousness is limited in what it can perceive. Long exposure to strange information, like things that defy logic, can cause mental instability, memory issues, and harm to mental health. So, limiting contact might help prevent these problems.

Reports of prolonged encounters often involve confusion, dissociation, obsession, and psychological distress. By contrast, brief encounters are dreamlike, symbolic, and less emotional, which tends to leave the experiencer shaken but able to carry on. This suggests an upper limit beyond which information transfer becomes damaging. Prolonged contact can overload mental processes.

Exposure might cause social instability, the spread of myths, or weaponization. In these cases, avoiding contact helps contain the effects.

Another constraint might be reciprocity. Consciousness-based interaction involves mutual influence. The longer the contact, the more they *become entangled*. Non-human intelligence could be affected by human emotions such as fear or aggression in unintended or unpredictable ways. Short interactions help prevent cross-contamination.

This model accounts for the fact that encounters are selective. People in altered states, like dreamers, meditators, trauma survivors, or psychedelic users, are already close to their perceptual limits. Interacting with them tends to be less disruptive overall, as it involves small, localized exchanges rather than widespread exposure.

This theory redefines avoidance as ethical restraint or necessity rather than secrecy. Like deep-sea creatures avoiding surface pressure and radiation, intelligences in different perceptual fields may need to adapt to their surroundings. *Staying outside these boundaries for too long could be harmful to both.*

Finally, the constraint hypothesis explains why intelligence agencies experience diminishing returns. Attempts to establish contact fail because they depend on conditions that cannot be improved. Increasing pressure undermines the very conditions necessary for interaction.

NHI is trying to protect the collapse of minds and systems that facilitate interaction in the first place.

Disclosure Delays

If we rely on human perception and mental limits, fully revealing contact with NHI is problematic because there's no clear way to present it. Any type of disclosure usually involves presenting something tangible, such as an object that can be observed, tested, and replicated. Encounters don't follow that pattern.

Governments are designed to release information that is verifiable and controllable. Experiences involving altered states, personal perception, or indirect communication don't conform to those rules. Even when radar data or physical traces are available, they seldom explain the entire event.

If the phenomenon is related to consciousness, it confuses the lines between science, psychology, religion, and intelligence. Mental health systems might need to rethink some experiences. Religious ideas could become more important, and intelligence agencies would face enormous challenges.

Public disclosure also raises expectations. People anticipate more transparent evidence and ongoing contact. However, increased attention and anxiety might actually reduce encounters, making evidence harder to gather even as demand grows.

From a risk-management perspective, partial disclosure is preferable to complete transparency. Ambiguity allows multiple interpretations without forcing a single explanation.

That's why disclosure stories tend to focus on specific aspects, such as craftsmanship, materials, or biological claims.

What's trickier is explaining the contact method. Interaction based on consciousness is hard to describe.

Ultimately, disclosure stalls because the phenomenon doesn't translate well to public explanation. The limit isn't solely technological; it's cognitive and institutional. Although hard to translate, the truth exists.

Inducing Altered States

Ayahuasca is a traditional Amazonian psychoactive brew used for centuries by Indigenous cultures for healing, spiritual insight, and communal rituals. It combines plants that render DMT orally active, producing powerful altered states when consumed in ceremonial settings led by trained shamans. In these cultures, ayahuasca is regarded as a sacred tool, and its use is guided by strict rituals.

Experiences often feature vivid imagery, emotional releases, physical purging, and interactions with what seem to be intelligent presences. Participants say they receive amazing insights, which supports the idea that consciousness might act more like a receiver than a creator of experience.

Modern research suggests that ayahuasca may help reduce depression and anxiety in some individuals and alter brain activity in ways similar to other psychedelics, although research remains limited.

Ayahuasca also poses serious risks, particularly from dangerous interactions with medications and pre-existing mental health conditions. Its legal status varies widely among countries, often restricted to religious or ceremonial use.

One common aspect of ayahuasca experiences is that people often report meeting autonomous entities that they see as teachers, healers, or guides. These encounters are described across many cultures and belief systems, and they usually feel meaningful, clear, and surprising.

Indigenous traditions regard these beings as real intelligences accessible by loosening perceptual filters, whereas psychological models treat them as manifestations of the unconscious. Neither explanation fully accounts for the consistency or informational quality of these experiences.

As a result, ayahuasca plays a key role in discussions of the limits of human awareness, linking shamanic traditions, neuroscience, and the question of whether the mind creates experience or filters it.

Chapter Twelve
Psychological Warfare

The Madness of MK-Ultra

The Cold War weaponized missiles, machines, and the *mind.*

By the early 1950s, U.S. intelligence agencies considered perception the next critical battlefield, one that surpassed land, sea, and air. Intelligence reports about Soviet psychological training, along with Korean War POW confessions, triggered a distinct fear: that the human mind could be manipulated by an outside force.

MK-Ultra was initially developed as a form of defence. However, subsequent events demonstrated how quickly defence can become an overreach when secrecy removes accountability.

The main idea of MK-Ultra was actually pretty straightforward: if we can change how someone thinks and perceives things, we can also influence their behaviour. This means it's possible to change a person's identity, erase or modify their memories. As a result, people's loyalty and trust could be controlled.

This wasn't just an abstract idea. Intelligence planners openly discussed "access states", the mental conditions in which resistance fades, and suggestions can pass through easily. Trauma, dissociation, and chemically altered states were treated as tools rather than problems. Pain was intentionally used as a technique.

Subjects were deliberately pushed beyond stress thresholds, operating on the premise that identity breaches would create entry points. As the mind detached from the body and memories muddled, a new state was unleashed, one characterized by increased openness, suggestibility, and compliance.

Psychedelics were used to unbalance the minds of individuals. LSD, mescaline, psilocybin, and other substances were administered to civilians, prisoners, and military personnel without their consent. The outcomes were unpredictable and frightening.

Some subjects became confused. Others became very suggestible. A few reported experiences that don't fit typical psychological categories: ego loss, supernatural contact, a confused sense of time, and encounters with NHI.

Researchers pressed forward. If trauma can break apart identity, and psychedelics can dissolve barriers, then the mind, without its usual filters, might access concealed information. Researchers discussed control, breaking through barriers, and defence mechanisms.

Dissociation Doorway

Dissociation, once seen as a way to protect against trauma, has been reinterpreted as a helpful trait.

Dissociative states may include:

- Disturbances in time perception (distortion)
- Incomplete recall of personal memories (dissociative amnesia)
- Unusual sense of identity (detachment from self)
- Derealization (detachment from reality)
- Altered bodily or spatial awareness
- Reduced integration of thoughts, emotions, and actions

These effects reflected the conditions sought in remote viewing and other consciousness studies to achieve optimal performance.

What MK-Ultra researchers failed or refused to acknowledge was that dissociation leaves scars. Splintering the mind isn't elegant because opening this door ripped the hinges off the frame.

Publicly, MK-Ultra was about mind control. In private, its goals were more sinister. The experiments sought to explore states of consciousness to control awareness.

Stargate viewed the mind as a receiver of consciousness, whereas MK-Ultra focused on what it could handle and the possible consequences.

Ethics Abandoned

Doctors, psychologists, and administrators knew that consent had been bypassed. Obtaining consent would create records, and records would lead to accountability. Subjects were chosen for their disposability.

When congressional hearings finally revealed MK-Ultra, the scariest part was more than cruelty; it was how normal it seemed. Memos spoke calmly about increasing doses, as if they were talking about everyday supplies. Deaths were ignored and downplayed. Files were regularly destroyed. Secrecy made the unthinkable seem normal and numbed everyone to the horror.

There's a deep institutional fear of the mind. MK-Ultra's influence can be seen today in interrogation methods, psychological operations, studies of dissociation and trauma, and the ethical rules governing human experiments.

If consciousness could be altered this much, it's obviously not in control. That realization might have been more harrowing than any foreign adversary.

The mind could be opened, but access came at a cost. The states that enabled intrusion also dismantled the structures that made information reliable.

Later programs stopped using force and instead focused on observing because causing harm was found to be unhelpful. Intelligence agencies realized—*too late*—that consciousness can't be controlled. If you break it, the signal disappears.

The Philadelphia Experiment

The *Philadelphia Experiment* is a legend from World War II. It claims that the U.S. Navy ship *USS Eldridge* was made invisible and briefly teleported during a secret military experiment in 1943. The story says the ship disappeared in a green haze, reappeared miles away, and then returned with crew members who were confused, injured, or even fused together.

The Philadelphia Experiment remains a mystery because it shows how easily perception can be affected when there's no clear explanation. Wartime sailors were put in situations they didn't understand, while strong currents moved through the ship. Lights appeared where they shouldn't, and compasses didn't work properly. The sailors reacted with nausea, confusion, and a feeling of being disconnected from themselves.

The Philadelphia Experiment legend stems from real

wartime secret technology efforts that were often misunderstood and exaggerated over time. Ships like the Eldridge were testing *degaussing and deperming*, which involved running high currents through coils around the hull to reduce the ship's magnetic signature and protect it from magnetic mines. To untrained observers, the sparks, humming sounds, and crew disorientation could seem mysterious or even supernatural.

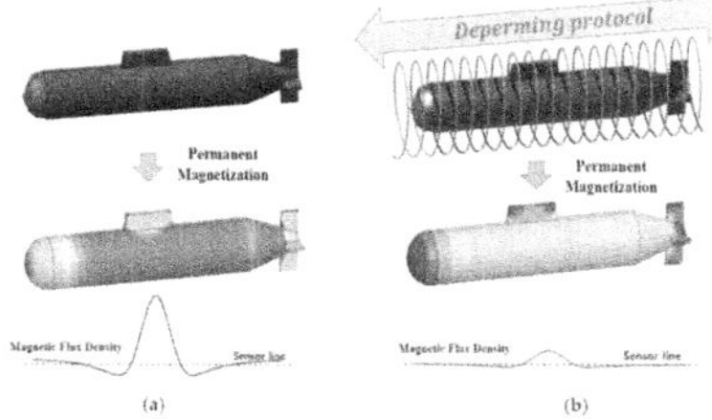

Schematic representation of ship deperming, showing how externally applied magnetic fields are used to neutralize residual hull magnetization.
Adapted from S.-H. Im et al., "Novel Deperming Protocols to Reduce Demagnetizing Time and Improve the Performance for the Magnetic Silence of Warships," *Energies* 14, no. 19 (2021): 6295. CC BY 4.0.

Early radar and detection systems sometimes gave false readings, making ships look displaced or duplicated on screens. Submarines tested soundproof coatings and stealth techniques to temporarily reduce their detectability. Sailors and observers who saw strange lights, unpredictable equipment behaviour, or crew members feeling ill often had no explanation. As these stories were shared, they were often *exaggerated* over time.

A hum or shimmer turned into a glowing field. Temporary confusion became teleportation. The goal of making a ship less detectable grew into the myth of invisibility. The pattern is clear: new technology, secrecy, and human imagination transformed simple scientific ideas into a legend that sounds like science fiction, even though it is based on real phenomena.

The Philadelphia Experiment shows how secrecy, confusion, and strange technology can influence beliefs more than facts. During war, governments used real secret missions and the fear of the unknown to control enemies and their own troops. A ship with weird electrical effects, broken instruments, and a confused crew was the perfect setting for rumours and myths.

Psychological warfare tricks people into doubting what is real, what technology an enemy has, and if they are safe. The Philadelphia Experiment was an excellent example showing how perception can be used as a weapon.

The Incident

In the early 1950s, researchers at *McGill University*, secretly funded by the CIA, conducted sensory-deprivation experiments under the supervision of psychologist *Donald Hebb*. Volunteers were placed in rooms where auditory, tactile, and visual input was minimized. Participants wore translucent goggles, remained still for hours or days, and were deprived of meaningful stimulation.

What was alarming was that within hours, subjects reported:

- Distorted perception of time
- Loss of coherent thought
- Vivid hallucinations
- The sense of a presence in the room

After prolonged exposure, many could no longer maintain

a stable ego. Some became highly suggestible, and a few completely broke down.

When regular stimulus is sufficiently dampened, the mind constructs its own reality.

Havana Syndrome

Beginning in 2016, U.S. and Canadian diplomats based in Havana began reporting sudden, localized episodes of distress. They started with a sharp sound, a pressure in the head, and a vibration felt. Then came the aftermath: dizziness, headaches, disorientation, memory issues, difficulty concentrating, and sleep disturbances. Some symptoms faded while others continued for quite some time.

The sensations appeared suddenly, often in specific rooms or locations, and sometimes ceased when the individual moved away. Medical evaluations later revealed signs consistent with mild brain injury in several cases, despite the absence of impact or visible trauma.

Early guesses pointed to directed energy or microwaves. Later, experts thought stress, expectations, or psychological factors might be involved. Official reports narrowed the possible causes, but some cases remained unexplained.

Havana Syndrome exposed a vulnerability from a new weapon. It showed that people experience real physiological effects even when there's no clear explanation for them.

Whether *Havana Syndrome* was caused by technology, the environment, psychology, or a mix of all three is less important than what happened afterward. The real harm came from the uncertainty, as those affected were caught between being believed and outright dismissed. The lack of explanation only made things more confusing.

In this sense, *Havana Syndrome* suggests that when the

nervous system becomes the battlefield, interpretation becomes difficult.

Project Artichoke

Project *Artichoke* began in the early Cold War as a darker form of interrogation. It asked, could a person's sense of identity be dismantled? Based on wartime psychological research and linked to what would become MK-Ultra, the program viewed the human mind as a system that could be broken down.

Subjects faced intense stressors like isolation, sleep deprivation, hypnosis, psychoactive drugs, and sensory disruptions.

Programs such as *Artichoke* and *MK-Ultra* revealed that identity isn't a continuous process. It depends on the stability provided by consistent sensory input from the environment. Remove enough of these elements, and the self is lost.

This idea changed the way people thought during the Cold War. Instead of using force, interrogators started confusing prisoners. Confusion worked better, and making someone feel isolated was even more effective than violence. When someone couldn't tell the time, where they were, or what was real, they didn't need to be forced; they gave up easily because they had nothing left to stand for.

Neuroscience confirmed that the brain uses prediction and pattern recognition to stay organized. When predictions go wrong, such as when signals arrive randomly, the nervous system struggles. Stress hormones flood the brain, memories get jumbled, and the part of the brain that helps us keep track of who we are fails.

Intelligence agencies found mind control wasn't like in movies, but something more dangerous: *corroding the mind.*

Sensory Deprivation

The *McGill sensory deprivation experiments* were important because they demonstrated what happens in the absence of stimulus. Within hours, the brain began generating internal signals to compensate for the loss of external input. Hallucinations were the mind's way of trying to create structure where there was none.

If perception happens inside us, what decides what we see or feel? Memory? Expectations? Culture? Or something even more basic about how we think?

The experiments were restructured. Sensory deprivation became a method for revealing latent processes usually masked by constant sensory input. The same logic would later be applied in float tanks, extended isolation studies, and astronaut training.

It turns out the void wasn't empty.

Induced Trauma

As Cold War programs grew, trauma was seen as a helpful tool. Dissociation, which used to be considered a mental problem, was found to be an ordinary response.

When someone becomes dissociated, they *feel disconnected*. From a mental viewpoint, this can be helpful. It allows for compartmentalizing feelings and denying certain things. A person might experience something without clearly recalling it, or remember it but not believe it was their own.

This idea reappeared in military survival training, interrogation resistance programs, and some therapeutic methods. The main difference was their goals: one aimed to protect, the other to gather information.

Credibility Problem

By the later stages of *MK-Ultra* and its satellite programs, a paradox became impossible to overlook. The more a subject's identity was undermined, the less reliable any information became.

Subjects in highly altered states talk freely, but their speech often drifts or isn't fully reliable. Their memories could be manipulated, and suggestions may influence what they remember. Although they could provide answers, they weren't very accurate. This led to a failure that couldn't be corrected with improved techniques.

This insight changed research from using force to observing, as in programs like Stargate. Instead of dealing with trauma, they focused on relaxation. Harm to the subjects was considered pointless and unethical.

Psychological Spillover

The legacy of these programs showed up in advertising, political messaging, interrogation doctrine, and psychological operations.

Today, modern psychological operations send mixed messages, overload attention, break trust, and undermine agreements. The aim is to exhaust people. When they can't tell what's true, it becomes easier to control them through confusion.

This approach closely resembles the outcomes observed in lab results. By removing coherence, introducing noise, and allowing the system to become erratic, it mimics the transition seen in MK-Ultra.

Destabilized Perception

Whether through sensory deprivation, trauma, psychedelics, electromagnetic speculation, or psychological stress, the outcome is the same. The mind loses its anchoring reference points and drifts.

In some instances, this is called a pathology.

This is why incidents like *Havana Syndrome* lead to institutional paralysis. Acknowledging unexplained neurological effects threatens to reopen questions that intelligence agencies have long attempted to suppress: how fragile the mind can be and how little control of it exists.

The main concern about *MK-Ultra* and *Artichoke* was that they demonstrated how a person's identity could be altered. If the sense of who we are can be altered by chemicals, psychology, or the environment, then our independence depends on those factors. How we see things can change without our realizing it, which makes consent unclear.

The lesson that intelligence agencies have learned, over time and at a human cost, is that accessing consciousness can lead to consequences. Any approach that bypasses the ego also eliminates the filters that keep our experiences stable, clear, and reliable.

Ultimately, psychological warfare revealed its limits: surrendering consciousness to control leads to confusion.

Chapter Thirteen

Deep Waters

I ventured out into the open ocean during a cruise through Alaska's Inside Passage. The ship started in Vancouver, travelling by sea toward Glacier Bay, with stops in Juneau, Skagway, and Ketchikan. For most of the trip, towering mountains stood nearby, which I found comforting.

Then, for a stretch, the land vanished.

As we sailed into deeper waters, farther from shore, the ocean's temperament changed as the surface grew restless. To me, the ship's motion felt like it was rotating on a pivot. I became terribly seasick for about five hours, as my body was overwhelmingly nauseous.

It was late evening, and the water was black, with no landmarks or lights except the ship itself. The waves were repetitive, a low hush that filled the night. Moonlight softened the darkness, casting a pale path across the water into the abyss.

With more than eighty percent of the ocean still unmapped, our understanding of what's beneath the surface is limited. Many parts of the ocean floor have never been directly explored, and entire ecosystems go unnoticed. People regularly travel through these waters confidently, even though we know less about them than we do about Mars.

Standing on that balcony, gazing into the distance, I'd imagined that moment would soon be lost in time.

We tend to see the unknown as something grand and distant, like looking up at the sky in search of mysteries beyond our grasp. Space and the ocean both feel perfect for that, far away and out of reach. We forget how little we know about the ocean, even though we live along the shores, because it's essential to our trade, military, and food needs.

Although I didn't get a sense of anything watching us, the environment made one thing pretty clear: whatever lay beneath the surface, whether biological, geological, or otherwise, could hide easily. As the land reappeared and the water calmed, that feeling faded, but the impression lingered. Maybe that's why we don't usually think of it as a place where NHI hides.

This chapter highlights the concept of proximity and suggests we may have overlooked questions about NHI, unidentified objects, and hidden underwater realms.

If unintelligent beings wanted to avoid contact, the deepest oceanic regions would be highly appealing, and we would continue sailing without even noticing them.

USOs: Definition and Historical Context

Unidentified Submerged Objects, or USOs, are entities observed moving beneath the surfaces of oceans, lakes, and other bodies of water that exhibit characteristics unlike those of any recognized submarines or marine creatures. USOs are often reported to accelerate to incredible speeds, change direction abruptly, and demonstrate trans-medium capabilities, seamlessly transitioning between water and air.

Right after World War II, the U.S. Navy began noticing unusual activity underwater during routine patrols and training drills. In 1947, off the Atlantic coast, sonar operators aboard a destroyer detected fast-moving, metallic objects making rapid vertical and horizontal movements.

These behaviours were unknown to submarines' abilities at the time. Although classified, logs indicate multiple witnesses confirmed these occurrences, describing objects that appeared to defy standard underwater physics.

* * *

In the 1950s, reports started to increase as submarine technology became more common. In 1952, an exercise in the Caribbean involved several U.S. destroyers tracking unknown submerged contacts that moved quickly, disappeared from sonar, and then reappeared miles away. Naval analysts later noted that these objects could maintain very high speeds at great depths, suggesting they might have had advanced propulsion systems.

* * *

In the 1960s, reports about encounters became more detailed and records improved. AFO sonar operators in the Pacific described seeing groups of glowing objects moving together over several kilometres. These objects were observed to climb and dive sharply while maintaining high speeds. These reports were part of Navy files later made available through FOIA requests, showing that trained military personnel regularly observed underwater objects moving along controlled, methodical paths.

During the Cold War, sightings increased, especially near key naval zones. Submarine crews and surface ships reported numerous encounters with unidentified underwater objects, some in groups, exhibiting trans-medium behaviour, surfacing briefly before disappearing into the air or quickly diving to depth.

Since 2000, AARO briefings and Navy records indicate that USOs are still being detected, primarily in areas such as East Coast training ranges, the Caribbean, and the Pacific. These objects can move quickly between water and air, emerging from deep water, then vanishing from sensors' view before reappearing elsewhere. Witnesses include sonar operators, pilots, and ship crews, providing multiple confirmations of these sightings.

Declassified Reports

Atlantic Frigate Sonar Incident

- *Witness Type:* Sonar operators aboard a U.S. Navy frigate
- *Object Behaviour:* Cylindrical object moving over 200 knots; sharp vertical and horizontal turns; glowing faintly on sonar; rapid acceleration/deceleration
- *Source Citation:* Declassified Navy FOIA report (Atlantic Fleet, 1997)
- *Note:* Demonstrates high-speed underwater maneuvering beyond known technology.

During an Atlantic encounter, sonar operators on a U.S. Navy frigate detected a cylindrical object moving at over 200 knots. The object suddenly changed depth and made sharp vertical and horizontal turns that no normal submarine could do. They described it as "smooth, metallic, and faintly glowing on sonar," with such precise movements that it seemed controlled by intelligence.

* * *

East Coast Training Range – Trans-Medium Object

- *Witness Type:* Fighter pilots and ship crew

- *Object Behaviour*: Emerged from below the water and rose into the air within seconds; radar confirmed vertical displacement; optical sensors captured a blurry reflective silhouette
- *Source Citation*: AARO briefing, 2019
- *Note*: Objects exhibited trans-medium capabilities, moving seamlessly from water to air.

Near the East Coast training range, fighter pilots observed an object that first seemed submerged, then quickly emerged and ascended into the sky. Radar tracked its swift vertical ascent, and optical sensors recorded a blurry, reflective shape. Crew on the nearby ship noted disturbed waters along the object's path, suggesting displacement greater than ordinary marine traffic.

* * *

Caribbean Destroyer Exercise

- *Witness Type*: Sonar-equipped destroyer crews (two ships)
- *Object Behaviour*: Coordinated movement of multiple objects; extreme accelerations and turns; intermittent disappearance from sonar and radar, reappearing miles away
- *Source Citation*: FOIA-release Navy exercise logs (2006–2007)
- *Note*: Shows coordinated, apparently intelligent movement underwater.

The Caribbean has seen reports of strange underwater objects. During a joint naval exercise, two sonar-equipped destroyers detected objects moving in coordinated ways beneath the water. The operators recorded accurate positions showing sudden accelerations and turns that normal technology can't explain. Sometimes, the objects seemed to disappear from sonar and radar completely, only to reappear miles away a moment later.

* * *

Pacific Submarine & Surface Vessel Observation

- *Witness Type:* Submarine crew and accompanying surface vessel crew
- *Object Behaviour:* Descended to depths over 3,000 feet while maintaining speed inconsistent with known submarine capabilities; corroborated by multiple sensors
- *Source Citation:* AARO/DoD report briefing, 2020
- *Note:* Confirms cross-platform corroboration and persistent deep-water presence.

AARO briefings mention incidents in the Pacific in which submarine crews and surface ships observed unknown underwater objects. Witnesses reported objects descending to 3,000 feet or more and moving so fast they seemed impossible. Multiple sensors confirmed the presence of these objects, and officers noted that witnesses on different vessels gave similar accounts.

* * *

These examples show a pattern: trained observers, such as pilots, sonar operators, and naval crews, often report underwater objects that move smoothly, traverse diverse environments, and appear intelligent. Whether these are new hi-tech devices, unknown natural phenomena, or something else, the records indicate that the oceans have long been sites of mysterious encounters.

The oceans are a huge hiding place. It's difficult to observe everything due to environmental and operational obstacles. Deep underwater, pressure and darkness make it hard for both people and instruments to monitor. Acoustic detection also has its challenges, such as signal attenuation, blind spots, and occasional false readings.

Civilian Reports

Fishermen across the Caribbean, Atlantic, and Pacific report seeing lights darting swiftly beneath the waves or objects slicing through the water at incredible speeds, leaving no wakes typical of regular boats. Private yacht crews and cruise ship passengers sometimes notice sudden disturbances in usually calm waters, such as large wakes or quick-moving shapes just below the surface, before the anomalies disappear as suddenly as they appeared.

While these accounts may not be as scientifically thorough as naval reports, they are similar to expert observations, suggesting this might be a widespread phenomenon.

* * *

Researchers have studied civilian experiences alongside military encounters to get a fuller picture. Jacques Vallée, in books such as *Dimensions* and *Passport to Magonia*, shared reports of underwater anomalies, along with sky and ground sightings, pointing to patterns of intelligent movement. Leslie Kean, in *UFOs: Generals, Pilots, and Government Officials Go on the Record*, mentions naval officers whose observations align with civilian reports, showing that these phenomena occur in both official settings and everyday life.

Contemporary databases such as *Grokipedia* and *UAPCheck.com* compile civilian and military reports, offering searchable collections of sightings, sonar traces, and optical observations. These sources show that, despite the oceans' vastness and ambiguity, the number of recorded encounters, both military and civilian, indicates a persistent, globally widespread phenomenon.

Including civilian reports alongside declassified naval records shows that USOs are not just limited to rare military events. Whether spotted by experienced sonar operators or a fisherman off Florida's coast, these objects behave consistently. Civilian observations help expand official records, offering a broader view of the oceans as places where strange phenomena occur.

Anomalies

Not all mysteries under the ocean reveal themselves through tracking or searching. Some unresolved curiosities lie on the seafloor or echo through the water.

* * *

In 2023, a NOAA expedition exploring the deep seafloor of the Gulf of Alaska came across something surprising that even experienced researchers found intriguing. At over 3,000 meters deep, remotely operated vehicles spotted a smooth, golden-colored sphere sitting alone on the seabed. It looked undamaged, shiny, and oddly shaped. When examined, it ruptured and released organic material, but scientists still had difficulty identifying it. It was recorded, discussed, and eventually considered biological, though there was no clear agreement on exactly what it was.

The deeper the ocean, the more often such peculiarities are found.

Sonar mapping has found large areas of the seafloor with long, straight scars. These scars sometimes run parallel for kilometres or cross each other at right angles. In shallow water, they are often caused by trawling, laying cables, or geological faults. But similar patterns have also been seen at greater depths where there is no known human activity or infrastructure.

In other places, sonar scans have found strange underwater structures that are hard to identify. They show rounded shapes, stepped formations, and sudden changes in height with clear edges. The Baltic Sea anomaly is a popular example: a circular object about sixty meters wide, found during a commercial sonar scan. Divers reported that their equipment malfunctioned near the site, but this remains debated. There are geological explanations, but no clear match nearby. As with many underwater mysteries, research slowed once there was no urgent threat or financial interest.

Even our most reliable way of sensing the deep sound still has enigmas. In 1997, hydrophones operated by the *U.S. National Oceanic and Atmospheric Administration* picked up a very low-frequency signal so strong that it could be heard across thousands of kilometres of the Pacific Ocean. This was called "*the Bloop*," and it was louder than any known marine animal. Later, it was believed to be caused by icequakes, but some strange sounds; *Upsweep*, *Whistle*, and *Julia* still can't be clearly explained. Some happen repeatedly, while others just once and then vanish.

Sound travels farther underwater than in any other environment. When an object moves, it produces an acoustic signature. Large objects operating at depth are unlikely to move quietly. However, the ocean is laden with signals that we can detect but struggle to interpret accurately.

* * *

There are places where oddities are due to their shape. Off Japan, near Yonaguni Island, large stone formations are underwater just below the surface. These formations have sharp angles, terraces, and flat surfaces that look intentionally crafted, but skeptics say natural cracking could create similar shapes. The debate is ongoing, and numerous people interpret these features differently based on what they think is possible and when.

These submerged structures pose an interesting question: are they artifacts of human creation, and does the ocean contain records of histories we have failed to discover?

What unites these cases could be described as liminal spaces. The phenomenon doesn't fit into known categories yet hasn't sparked any investigations.

If intelligence, human or otherwise, wanted to observe without being seen, the ocean would provide plenty of hiding spots. We look at the stars because they seem out of reach, but the ocean might be a better cover.

Part Five

NON-HUMAN INTELLIGENCE

Chapter Fourteen

Nature or Intelligent Design

R*ichard Feynman* once explained something that initially seems counterintuitive. Most of a tree's mass doesn't come from the ground. It comes from the air. Trunks, branches, and rings of wood laid down over decades are built primarily from carbon dioxide, an invisible gas.

Through photosynthesis, a tree absorbs carbon from the atmosphere and, using energy from sunlight, synthesizes sugars and cellulose. The soil contributes water and trace minerals, but the bulk of what we recognize as "tree" was once empty space.

Dynamic Life Cycle

Feynman vividly described the process: sunlight strips oxygen from water. The Sun's energy splits water molecules, freeing hydrogen to bond with carbon dioxide (CO_2) to form methane, while oxygen is released into the atmosphere. In this way, light, air, and water are transformed into solid matter. We instinctively view growth as soil consumption, but it is, in fact, a process of transformation. The tree is *a standing wave* within the environmental flow. *Form born from flux,* and once you understand this, it becomes harder to see nature as passive.

A tree gathers energy, filters what it receives, and shapes its form. Plants have been shown to detect vibrations in their surroundings and respond selectively.

When caterpillars nibble on leaves, the vibrations they produce are different from those caused by wind or rain. Plants can tell the difference and only ramp up their defences when the signal shows there's a real danger. While plants don't hear as we do, they do detect, interpret, and react to important information.

Vibration is a language that nature speaks fluently.

Insects constantly navigate this vibrational world. Many detect airborne sound through tiny tympanal organs, which are small eardrums located on the legs, thorax, or abdomen. For example, moths can hear bats' ultrasonic calls and respond immediately with evasive manoeuvres.

Echolocation

Others detect vibrations through solid surfaces. Leaves, stems, and flowers serve as transmission lines, conveying signals between insects. Courtship, territorial warnings, and threat detection all occur via oscillations that travel through plant tissue.

What's even more remarkable is how insects integrate multiple invisible signals simultaneously. Bees become positively charged while flying due to friction with the air. Flowers, anchored by their stems and roots, tend to carry a negative charge, creating a subtle electric field around the plant.

When a bee approaches, pollen grains can jump across tiny air gaps, drawn by electrostatic forces. Bees can also detect subtle shifts in a flower's electric field, helping them determine whether it has been recently visited and has less nectar.

Overall, ecosystems are complex systems that constantly exchange signals through electrical, vibrational, chemical, and acoustic means.

Birds also play a part in this unseen network. When they flap their wings, they produce familiar sounds and vibrations. Their calls help shape the acoustic environment, influencing how other animals communicate and react. Interestingly, insects have learned to recognize the sound of birds flying as a warning, even if the birds aren't trying to send a message.

Nature is overflowing with unseen messages, signals that weren't intentionally sent to communicate, yet they persist and speak to us all the same.

The Carrington Event

In 1859, during what is now called the *Carrington Event,* a massive solar flare emitted a wave of charged particles towards Earth, and when it arrived, Earth's magnetic field shook violently.

Telegraph operators across North America and Europe reported something extraordinary: *telegraph systems continued transmitting messages even after their batteries were disconnected.* Some operators were astonished as paper caught fire, and long lengths of wire, acting like enormous antennas, had electricity induced directly into them by the changing geomagnetic field.

Even though nothing was connected, the system still responded to environmental anomalies.

This is explained by Faraday's law: a changing magnetic field induces an electric current in a conductor. Telegraph wires didn't store energy; they converted changes in the magnetic field around them into power. The Sun's magnetic field affected the electrons, causing them to move and triggering a response.

At every scale, whether it's leaves and insects, forests, or telegraph networks, nature is remarkably sensitive to unseen forces. Electric fields, vibrations, pressure changes, and magnetic fluctuations serve as inputs. Life has evolved to endure and harness them.

If natural systems already detect, integrate, and respond to subtle signals across many areas, it becomes harder to draw a clear line between what's a "*natural process*" and what's a "designed system".

If that's true, intelligence could access natural channels through which information flows to communicate. Nature was meant to be *understood* by something able to comprehend it.

Mycelium Intelligence

Beneath the forests, there's an invisible team working together. Mycorrhizal fungi spread through the soil, linking tree roots into big, lively networks that can reach far and wide.

These connections help nutrients travel to where they're needed, send signals to warn about dangers, and allow healthier trees to help those in trouble. It's nature's secret, cooperative internet at work.

The system manages itself through closeness, pressure, and reactions. The result is a shared awareness: the forest collectively responding to drought, disease, or plenty. This intelligence isn't thinking or planning. It's a *pattern of relationships*, an ongoing process shaped by sensitivity to the environment.

The fungal network isn't sending messages like wires do, with information flowing directly from one point to another; it changes in response to differences. When one part changes, it influences others, and the entire system adjusts. So the forest is a living medium, and life learns to understand these changes, since survival depends on them. Sensitivity transforms into a form of understanding.

Fields are more than background; they actively contribute. Chemical scents, electric charges, sound waves, and magnetic changes affect behaviour long before anyone consciously notices. *Survivors* are those who tune in to these hidden signals and adapt to their environment. Intelligence would be the ability to stay responsive to a constant stream of signals.

As *Darwin* is often paraphrased, survival favours not the strongest or the smartest, but the most adaptable to change.

Any clever entity wanting to connect wouldn't need to create new ways. They just need to learn how to communicate through the ones already there.

Magnetoreception & Field Navigation

Many species navigate using signals outside human detection. Birds, sea turtles, and even single-celled bacteria possess magnetoreception—the ability to detect Earth's magnetic field and use it for orientation and migration. Birds have a special protein in their eyes called Cry4, a type of cryptochrome that is sensitive to blue light. This helps them see magnetic fields, probably as a visual overlay, making navigation easier.

These organisms respond to landmarks and memory but also to a planetary-scale magnetic field, which they interpret as navigational data. The magnetic field becomes a map, and the planet a warning system. What appears as empty space provides precise guidance for actively interpreting Earth's fields.

Cells in living tissue react to signals automatically. They sense tiny bioelectric fields, currents in membranes and nearby cells, and respond by deciding where to grow, when to divide, and how to shape themselves. Researchers at *Tufts* and *Harvard* have found ways to help entire limbs and organs regenerate by guiding these invisible patterns.

Morphogenesis is a dynamic process in which everything responds and changes around us. It's a bit like forests and fungus networks we've seen, creatures that can sense magnetic fields, and ecosystems that adapt to unseen signals.

Intelligent Design

During development, cells organize into organs and the body without a central blueprint. Each cell responds only to local chemical gradients, mechanical pressures, and electrical cues, yet coherent forms emerge with remarkable reliability.

This process—*morphogenesis*—shows that order depends on recognizing constraints and feedback. The system is designed to respond intelligently to the information it has at any moment. In this way, design comes to life through interaction.

Slime moulds and ant colonies add an interesting layer to

the story. While each one is simple on its own, together they solve problems that would challenge a single mind. Slime moulds can find their way through mazes, predict cycles of change, and reroute resources across their ever-changing bodies. Without a single intelligence guiding them, ants and bees organize entire societies, dividing up work, defending their territory, and finding the shortest routes between food sources and their nests.

Plants communicate in ways we are only starting to understand. Roots investigate the soil, sensing chemicals and electrical signals. The Venus flytrap detects vibrations and closes only when it recognizes a specific pattern that signals the presence of prey. Fluids and hormones circulate through plants in response to touch, light, and moisture. These organisms are capable of thinking, making choices, and remembering, all of which are driven by internal processes.

Birds, turtles, and bacteria can detect signals over large areas. Earth's magnetic field helps them find their way with a hidden map of direction. The invisible holds important details for those who can pick up on it.

Faceless Intelligence

A form of what we call intelligence is nature's way of listening. Forests, fungi, cells, insects, birds, and turtles all read, respond to, and adapt to fields and signals that are barely noticeable to us. Life has been practicing this language for billions of years, long before humans began to understand it.

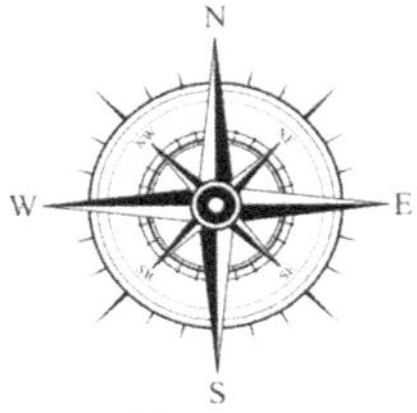

Field Navigation

The Sun's magnetic field extends across the planet. Chemical and electrical signals ripple through soil and tissue, shaping forms, guiding movement, and supporting survival.

If an outsider wanted to talk to us, they could just use the existing signals, currents, vibrations, and fields. Some lifeforms understood all along.

Intelligence is a part of life, present in everything around us. It flows through forests, fungi, cells, roots, and wings, and reaches beyond Earth, following magnetic lines, sailing on solar winds, and connecting the vast fields between stars.

The stars and galaxies are parts of the universe and active players in the emergence of awareness. Life becomes part of a larger consciousness that reaches across worlds, suns, and the empty spaces between. It is big, old, and faceless.

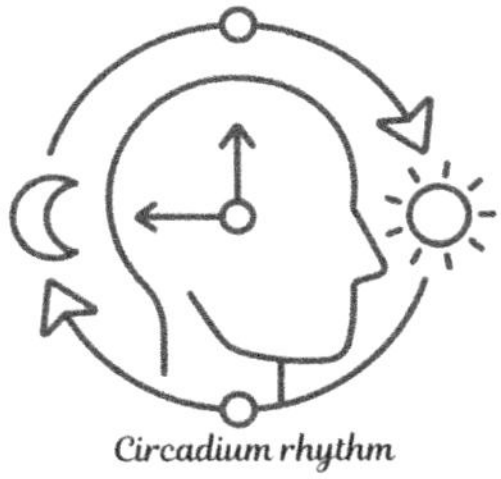

Circadium rhythm

Modern biology reveals that life depends on more than just local information. Cells maintain their rhythms in darkness, and organisms communicate over long distances. Physiological processes remain linked to larger cycles that extend beyond their immediate surroundings. These patterns continue even without obvious cues, illustrating that life stays aligned with broader rhythms.

Circadian biology exemplifies this idea. Internal clocks control sleep, hormones, immune responses, and metabolism with high accuracy. While these clocks function independently, they slowly drift and need to be realigned using environmental cues. Their adjustment isn't driven only by light but also by solar activity, atmospheric changes, and geomagnetic variations.

* * *

Disturbances in Earth's magnetic field have been linked to changes in heart rate variability, cognitive performance, and mood. The links are modest but consistent. Life is tuned to planetary conditions in ways that operate below conscious awareness.

This alignment extends outward. Cosmic radiation, coming from distant space events, constantly interacts with Earth's atmosphere and life forms. Over long periods, changes in this radiation can affect mutation rates and genetic diversity. Evolution, often seen as a process happening only on Earth, is actually influenced by events happening far beyond our planet. In other words, biology develops within a *galaxy-wide* context.

Systems respond to their surroundings the way rivers carve paths that follow the land. Crystals reflect boundaries and record information about their environment. Living systems do the same, but with greater sensitivity and a need for feedback to stay organized.

This connection is really important. The same basic ideas that help cells grow, like electric forces, energy use, and field interactions, also explain stars, galaxies, and space plasma.

Magnetic threads stretch across the universe, and biology builds on this idea.

From this view, *intelligence isn't just a rare gift but something that arises when systems receive continuous information.* As they get more Feedback, they become more responsive. When responsiveness improves, systems start to coordinate. Over time, this coordination begins to look like awareness, often enough to suggest that intelligence isn't just a fluke.

Emergent Intelligence

Nature and design are intrinsically linked. Nature maintains the balance and order of our world, while design translates this order into tangible forms over time. What we consider intelligence are systems that, given enough time, have learned to remain sensitive and adaptable without breaking down.

* * *

Trees grow from the air. Cells follow electrical signals. Animals find their way using planetary fields. Life changes under the influence of the cosmos.

What we're gradually noticing is a capability and power for matter to become conscious. Once this awareness develops, the universe begins to observe itself.

Chemically Generous Bennu

In 2023, NASA's *OSIRIS-REx* mission returned a small capsule containing material from the near-Earth asteroid *101955 Bennu*. Bennu is a carbon-rich remnant from the earliest days of the solar system, a long-destroyed parent body that formed more than 4.5 billion years ago. Unlike planets, which have been melted, reprocessed, and recycled by geological activity, Bennu has remained largely unchanged since then. It is, in effect, a preserved sample of primordial chemistry.

Samples brought back to Earth show they contain many of the key ingredients for life: at least fourteen amino acids, all five DNA and RNA bases (A, G, C, U, T), sugars like ribose, phosphate minerals, and water-rich clays such as serpentine.

These materials formed long before Earth existed, through natural chemical processes in space and within Bennu's parent body, which was destroyed long ago. There, liquid water once interacted with rock. In other words, the basic ingredients for life were already forming in the early solar system, even before any life existed here.

What's truly compelling is that Bennu maintains its chemistry in a pure, unaltered form. The presence of both left- and right-handed amino acids, along with organic molecules attached to mineral surfaces, suggests a natural mechanism that could facilitate their gathering and organization.

This supports the idea that early Earth was probably a water-rich world in which asteroid-delivered organics accumulated in oceans and hydrothermal zones. In these environments, energy gradients, mineral pores, and water-rock interactions might have promoted the formation of more complex structures, potentially giving rise to RNA-based systems before the appearance of true cells.

If Bennu can do this, it's likely that other carbon-rich asteroids and planetary systems can too, since the same physics and chemistry happen wherever stars are born. This means the chemistry for life might be common across the universe, even if actual life is rare. Planets could have a molecular advantage, thanks to ancient stardust and asteroid chemistry.

Bennu suggests the universe can support life wherever conditions are right. The real mystery is how often and under what circumstances they form something truly alive.

Molecules began forming longer chains resembling early RNA, which were crucial for supporting other molecules and occasionally copying themselves. Once copying began, natural selection took over, and chemistry regenerated what worked best.

Eventually, fatty molecules assembled into simple bubbles that enclosed reactions, establishing an inside and outside. These proto-cells weren't alive as we understand today, but they were stable enough to grow, evolve, and compete. Over time, this progression resulted in the development of true cells, DNA, proteins, and the biological systems we are familiar with.

Here's the main idea: space provided many of the ingredients for life, but Earth was where those ingredients finally came together to form living things. This suggests that the chemistry of life might be quite common throughout the universe, even if actual living creatures are rare.

Chapter Fifteen

Encrypted Geometry

In various ancient civilizations that never had contact with one another, you can still find familiar shapes like spirals, circles, triangles, eyes, and grids. These appear in temples and cave paintings. They help brighten up spaces and preserve history.

Throughout periods of historical chaos, such as religious cleansings, imperial expansions, and political changes, people have often tried to destroy anything that questions authority. Fortunately, some truths are more resilient, hiding in customs, symbols, and rituals.

Secret societies openly display their symbols because they're not meant to be private. Just like a message hidden in the angles of a cathedral, the proportions of a mandala, or the steps of a ritual, meaning becomes part of the design itself.

Geometric knowledge is safe through history's chaos. Initiations and rites serve as mental filters, helping people learn to see the important patterns. Only those who are in tune with the design can uncover the meaning inside.

Archetypes

The *serpent* winds around themes of renewal and secret knowledge; its *eye* watches over both as a witness and a wise guide. The *tree* reaches between different worlds, the *spiral* shows time and cycles, and the *labyrinth* stands for journey and coming back home.

Myths are more than just stories; they serve as helpful treasures, like cognitive fossils that preserve important truths across many generations. They carry knowledge strong enough to cross different cultures but remain unnoticed by the censor.

Those who uphold these patterns are guardians of what is too powerful to be publicly disclosed. Priesthoods, scribes, monastic orders, fraternal societies, and even modern intelligence agencies have all served in this role.

Civilizations may forget some details, but they usually remember how truths can be hidden. A cathedral can become a star map, a ritual might turn a diagram of the universe. Forbidden knowledge becomes part of culture.

Pattern ~ Craft

In many modern reports of UAPs, as well as in older mystical or visionary stories, people often first notice a pattern before a craft. Lights forming grids, triangles, or spirals are common symbols that appear in dreams or altered states. The experience usually starts with recognizing geometric shapes in the sky or just behind the eyes before any real object appears.

From a neurological standpoint, our visual system initially detects edges, symmetry, and motion prior to recognizing objects. Geometry serves as the fundamental framework for perception. Employing a pattern that captures attention without overwhelming us is the safest way to begin understanding.

Geometry acts like a mental handshake, checking if we're open to contact before physical signals arrive.

A physical object has cultural meaning: we see it as a machine, a threat, or technology. This makes it easy for us to see it as intentional without needing further interpretation. A triangle of lights in the sky tells us something is purposefully organized. That alone can lead us to recognize it as intentional without requiring interpretation.

Some researchers studying close encounters have noticed a familiar paradigm in reports: first, strange lights or shapes; then, a presence; and only afterward does an object appear. This suggests that what appears to be a vehicle might be the final stage of a process that begins with our perception of it. If that's the case, then geometry is the first thing we'd notice.

From an NHI perspective, this makes sense. Geometry is one of the few concepts shared by humans and physics. Circles, ratios, symmetry, and repetition are found in atoms, orbits, and waveforms as well as in temple drawings. If an intelligence wanted to communicate without written language, a pattern would be the simplest and clearest way to do so, with meaning kept to a minimum.

This also explains why people often see sigils, grids, or mandala-like shapes during altered states associated with UAP experiences. These might not be hallucinations in the usual sense, but rather the mind's interpretation of external stimuli. Essentially, the mind turns something hard to understand into a familiar pattern: geometry. Just as radio static becomes sound when tuned in, raw signals become recognizable shapes once perceived.

Mandala

Geometric patterns are a straightforward way to connect with others. Our perception often begins with how we see things, and geometry is the impression this process leaves. This suggests that UAP contact originates in the mind's pattern-recognition system, not in the sky.

Nature's Canvas

Fibonacci numbers pop up in petals, pinecones, and hurricanes. As you walk along a coastline, wander through a forest, or cross the dunes, you'll notice it: beautiful spirals in shells, branching trees, swirling galaxies above, and the gentle rhythm of waves. Fractals are everywhere, as if the universe is repeating a simple, lovely pattern at every size.

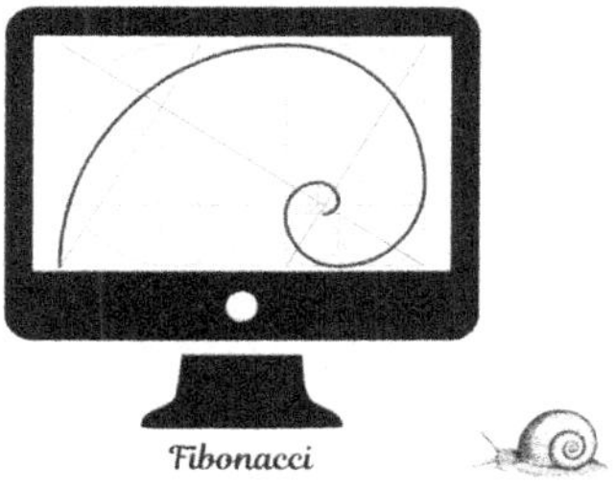

More than being pretty, these patterns play a real role in how things grow, move, and flow. Plus, our minds notice them right away, even when we aren't thinking about them.

NHI seeing nature as a canvas makes sense. Patterns like fractal trees, spiralling shells, and crystal structures might be natural ways for life to communicate. Humans might already be tuned into this. Nature acts as a messenger through patterns, repetition, and symmetry.

Think about what this means for contact. If an NHI wanted to reach out, it could do so by subtly *adding patterns to the world humans already see.* A spiral of lights in the night sky might look like a sunflower seed spiral. A quick triangle of glowing objects could resemble a crystal pattern or a river delta. Humans tend to notice these signs because our brains have evolved to recognize such patterns, which the intelligence might also use.

Popular culture has occasionally explored this idea. The 1977 film *Close Encounters of the Third Kind* imagines first contact unfolding through recurring tonal sequences and geometric light patterns, communication based on patterns.

This may help explain why UAP encounters often seem like a dream. They tend to match familiar geometry that we recognize on a subconscious level. Geometry serves as a common language embedded in the fabric of matter and our perception.

The underlying suggestion is that intelligence involves

recognizing order in unexpected places. Natural patterns could have directed human focus for millennia, priming us to perceive something larger and beyond human *nature*.

Crop Circles

Crop circles are quite a curious phenomenon. The landscape is used to express language. The medium is living matter, and the message, if there is one, is geometric: circles, spirals, lattices, fractals, and sometimes diagrams so exact they look like technical schematics. They really do illustrate a sense of *order*.

What makes them interesting is the mystery of how they express themselves. A crop circle is like a hidden treasure waiting to be found, just like a mandala or a fossil; it only makes sense to someone who notices it. This suggests that intelligence is trying to communicate to us through our cultivated crops, using symbols that aren't tied to any particular society.

Many formations mimic features found in nature: spirals like those of shells and radial symmetry like that of flowers. Others resemble mathematical objects, fractal iterations, nested ratios, or patterns that suggest wave interference and orbital motion.

Crop circles appear to draw on the same visual language as biology and physics. They demonstrate a combination of natural growth and abstract design.

If someone imagines NHI using nature as a canvas, crop circles become an almost logical choice. Fields are vast, uniform, and visible from above. Crops regrow, leaving no permanent damage.

This explains why conversations about hoaxes often miss the main point. Even if some formations are made by people, the real issue is that humans naturally see patterns in the natural world and assume they have a purpose. The true meaning of crop circles depends on who makes them and how people interpret them, since the patterns themselves are trying to communicate something.

Perhaps an NHI might avoid cultural symbols altogether, since they belong to us. Geometry, however, belongs to everything. Circles and spirals existed before the eyes could perceive them. Fractals shaped coastlines before we did.

NHI can remain anonymous by using nature as a medium, as it's trusted by humans. Crop circles act as a proof of concept, demonstrating that order can materialize where it wasn't expected and that perception becomes an integral part of the exchange.

Nazca Lines

Stretching across the Peruvian desert, the Nazca Lines are a striking collection of enormous animals, geometric patterns, and abstract figures visible only from above. When you stand on the ground, you'll notice faint paths; only from a higher viewpoint does their full meaning become clear. It's a bit like the crop circle phenomenon: the message is concealed in the world around us.

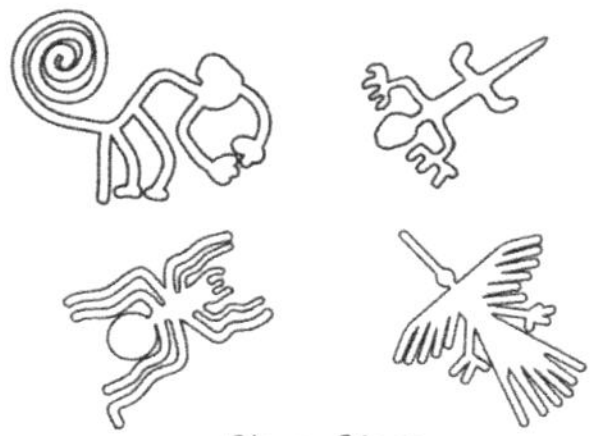

Nazca Lines

Their size indicates they weren't intended for ordinary observation. Some theories suggest that the creators aimed to signal the gods or spirits.

Geometry integrated into the landscape serves to make contact and to function as a beacon. Spiral, trapezoid, hummingbird, monkey. Every shape conveys ideas about ratio, proportion, movement, or connection. The desert floor is a canvas from the air, and the human mind is a receiver.

Like crop circles, the Nazca Lines show us that communication can be about patterns embedded in the environment that only those who notice can understand. These lines might also help ideas pass down through generations: one person sees the shapes, understands what they mean, shares that understanding, and the pattern keeps going, even if the original reason is forgotten.

Crop circles are temporary and ever-changing, whereas the Nazca Lines are large and permanent. Both prioritize patterns over detailed stories. They transcend language and are instantly recognizable, as they're designed to capture attention and engage the viewer's mind.

Viewed together, crop circles and the Nazca Lines use geometry as a form of communication.

Forbidden Knowledge

Forbidden knowledge can be risky because it challenges

authority and opens up new ways of understanding that don't always align with human thinking. This might include ideas about perception or physics that feel unfamiliar to the mind that's been shaped by culture and biology.

If NHI encodes information, it would likely do so in designs rather than clear statements, leaving the meaning obscured until a *suitable mind* can recognize it. This way, understanding grows gradually rather than all at once, helping both the message and the recipient avoid the risks of knowing too soon.

Ongoing Contact

Sigils and mandalas may function as condensed forms of communication. Visual algorithms that store and transmit information without language. Their strength lies in shaping perception. A single image can contain layered instructions that become clear through attention or repetition.

Archetypes

Archetypes reappear because these constructs regularly engage with human cognition, activating familiar patterns of attention, memory, and interpretation.

Using similar forms across different times and places shows how they influence one another. When a pattern always causes the same response, it remains powerful. This creates a cycle:

the image affects our perception, and our perception, in turn, strengthens the image.

Myths and rituals connect through actions that align with focus and feelings. They help create mental states where information can be received and stored. In this process, language isn't as important because meaning is understood more through configuration than through specific words.

If an NHI were to communicate with humans without using our language or biology, this approach would be both effective and prudent. It ensures information continues without overwhelming the receiver, revealing itself only when the recipient's cognitive state is appropriate. The communication occurs *indirectly*.

Encounters

This setup helps explain why UAP encounters often occur without much dialogue or instructions. Witnesses usually don't report being told anything in the usual way. Instead, they notice patterns such as sequences of lights, geometric shapes, heightened concentration, or sudden waves of understanding that are hard to put into words. The experience has a cohesion, but it's missing the language to describe it.

Our perception of an encounter is shaped by its form, leading us to be more open-minded instead of fixating on specifics. This could also explain why many encounters alter our experience, making time feel strange, evoking intense emotions, producing clear yet dreamlike images, or causing us to forget details quickly.

Our minds function like processors, retaining impressions even after the encounter ends. The observer becomes part of the interface, with meaning developing through participation.

Myths and ceremonies are more about preparation. They

help train the mind to stay calm and focused in uncertain situations. Cultures that keep these practices may have mental strategies that help them handle interactions without falling apart.

This highlights an important barrier: the intelligence here functions like a system that influences its surroundings. Contact marks a moment of alignment, and encounters are brief, recurring periods, much like a solar or lunar eclipse.

Chapter Sixteen

Contactees

Faith's Witness Account (name changed for privacy)

The following account was shared with me by a private witness who requested anonymity. Unlike highly publicized abduction narratives, her experiences involve brief recurring encounters with shadow-like figures. While lacking dramatic physical interaction, the consistency of the sightings and their similarity to reports from unrelated witnesses make the account analytically relevant.

The Rearview Figure

Faith was driving alone at night (2 am), the road empty, with only the low, familiar hum of the van's engine. What came first was a sudden fear that she was no longer alone. It was a feeling she had experienced many times before, most often while driving.

She checked the rearview mirror, and for around ten seconds, a dark figure remained there.

A tall, black silhouette sat behind her, sharply outlined yet strangely dense, as if made of a shadow in the darkness. The shape was definitely humanoid. The hair appeared full, puffing outward, giving the figure a defined, almost deliberate outline. It didn't move or react, just sat in the seat behind her, as if aware of being watched.

Her heart pounded as she turned to look directly behind her. The van was empty, and the encounter ended as abruptly as it had begun.

She reports that this was not an isolated incident. Over the years, she has repeatedly noticed the same figure at the edges of her vision, brief impressions that vanish the moment she turns to look at them. These sightings occur most often when she is in motion, especially while driving, a detail that appears in many other accounts investigated for this study.

At other times, she remembers, the shapes vary. Low to the ground, they take the form of a canine. She calls them shadow dogs; they're dark outlines that cross her peripheral vision and vanish before she can see them clearly. Similar animal-like silhouettes have been reported by other witnesses, often in the same fleeting, peripheral manner.

In these accounts, a few things tend to stand out: there's no sound, no direct contact, brief sightings, and a persistent feeling of being watched rather than approached. It feels like the experience lasts just long enough for the witness to notice a presence, then it fades away, leaving no physical traces, just this lingering sense that it wasn't imagination.

This account matches a common pattern seen in many witness reports. There is neither sound nor physical contact, but the feeling of being watched remains consistent.

* * *

A quick internet search reveals that similar reports often describe humanoid or animal-like shadows that briefly appear without disturbing the environment. These features suggest a phenomenon on the edge of awareness, manifesting just enough to avoid being clearly recorded.

Private Witness Case: Repeated Orange Light Sightings (Autumn 1998)

The following account was shared with me directly by a private witness and his parents during a conversation. Each family member independently recalls the same series of events from the autumn of 1998. Unlike classic abduction narratives, the case involves no missing time, physical contact, or encounters with humanoids. Its significance lies in repeated observation over multiple weeks, multiple witnesses within the same household, and the consistency with which the object's movement and disappearance were described.

Orange sphere

Between late September and Halloween of 1998, the family saw a strange orange light over their neighbourhood five times. The father was there for all the sightings, the son saw most of them, and the mother saw at least two. Their descriptions were the same, even years later: a shiny orange sphere that moved in ways normal planes can't, and its disappearance seemed deliberate.

The initial sighting happened late in the evening when the father saw an orange ball of light high in the sky, well above the trees. At first, it appeared to move steadily, travelling along a street before suddenly changing direction. Its movement didn't match that of a typical aircraft or satellite. As he observed, the light suddenly flared, intensifying sharply, then expanded both upward and downward, creating a brief cross-shaped flash. Almost immediately, it darted sideways at incredible speed and vanished.

Later that night, both the father and son saw the light again. Recognizing it from before, they decided to follow it on foot. The object moved slowly enough for them to track it as they walked down the street towards a nearby park. It appeared to hover above the trees, maintaining a steady altitude. When they reached a point where they believed they would have a clearer view, the light disappeared without warning.

Another evening, the mother observed the phenomenon for the first time. She saw what she described as a silver, reflective object in the sky and called out for the others to bring binoculars. When they joined her, all three could see it. As they watched, the object did not blink out or speed away. Instead, it seemed to fade gradually in a controlled manner. The mother described it as a window or doorway closing, as if a curtain were being drawn from one side to the other.

The father later recounted a similar event, likely from the same night. He described the light's disappearance as resembling a solar eclipse, in the way it moved. The light appeared to close in from one side to the other, rather than gradually shrinking or dimming evenly. The son remembers observing the light move back and forth across the sky before it disappeared.

Some time later, the family spoke with a close family friend and her teenage son. The teen described seeing the same orange light several blocks away. He and his mother watched it as they walked along the road toward a nearby school, where it seemed to hover over the area before vanishing. That was their only encounter, but the description closely matched what the family had already seen.

Halloween marked the last sighting. After that night, the object did not return. While none of the witnesses could identify what it was, they were consistent in their observations: the colour, the controlled movement, the sense that it reacted to being watched, and, most strikingly, the manner of its disappearance, as if something had been slowly shut.

* * *

Few subjects in ufology reveal the tension among belief, psychology, and perception more clearly than the testimonies of contactees. These are individuals who claim direct interaction with non-human intelligences, sometimes physical, sometimes psychological, or in between.

Their stories range from benevolent encounters with radiant beings who impart spiritual wisdom to terrifying accounts of forced medical examinations aboard strange craft. Across decades, the narratives change drastically, yet certain motifs remain consistent: missing time, humanoid entities, telepathic communication, altered consciousness, and a lasting sense that reality has somehow warped.

* * *

During the 1950s, stories emerged about cosmic teachers warning humanity about nuclear dangers. Later in the twentieth century, narratives centred on invasive visitors, trauma, and existential uncertainty. Whether these stories involve extraterrestrial beings, altered states of consciousness, psychological projections, or a mix of all three remains a matter of debate.

Analytically, what matters is that the experiences are real to those who report them. Their emotional consequences are so profound that relationships break down, and their worldviews collapse. People become convinced that they've crossed a boundary most others never encounter. The mind struggles to make sense of contact with an unknown phenomenon.

Orfeo Angelucci and the First Contactees

In the early 1950s, during the first major wave of UFO fascination, *Orfeo Angelucci* was known as one of the most influential "contactees." He was part of the same cultural moment as figures like *George Adamski,* when flying saucers were thought of as signs of spiritual evolution. This was the height of Cold War anxiety, when nuclear fear mingled with religious longing and postwar utopianism. Against that backdrop, *Angelucci's* encounters felt almost prophetic.

He talked about repeated encounters with friendly, human-like, often glowing or Nordic-looking alien beings who communicated with him verbally and through telepathy. These beings took him on their ships, showed him visions of other worlds, and shared moral and cosmic lessons. They warned humans about nuclear weapons, environmental damage, and losing spirituality. Instead of fear, *Angelucci* saw it as an initiation. He felt chosen and trusted with special knowledge.

His 1955 book, *The Secret of the Saucers*, came across more like a spiritual story than a serious investigation. There are no surgical tables, invasive procedures, or terrifying stories about missing time. Instead, it felt like a modern religious experience, but with futuristic imagery. Many scholars see *Angelucci* as a spiritual mystic who happened to incorporate spacecraft into his worldview.

This difference is important. His stories predate the horror themes that later became common in abduction stories. From spiritual "space brothers" to unfriendly visitors, these examples show that contact stories change over time and across cultures.

Betty and Barney Hill

Angelucci represents the spiritual beginning of the contactee era, but *Betty and Barney Hill* represent the beginning of modern abduction mythology. Their 1961 case became the template for countless later reports.

While driving home through rural New Hampshire, the *Hills* reported seeing a strange object following their car. What began as a UFO sighting became something far stranger when they later realized they had lost hours, with no clear memory of what had happened. Both experienced anxiety and recurring distress, particularly *Barney*, whose fear was intense and immediate.

Under hypnosis, they remembered being taken aboard a craft by humanoid beings and being subjected to medical exams. *Betty* described a star map shown to her by one of the entities, a detail that became well known in UFO circles. The beings seemed clinical rather than spiritual, and the encounter felt invasive.

This case established the core architecture of the modern abduction narrative: missing time, medical procedures, psychological trauma, and non-consensual contact. It also introduced one of the field's greatest methodological problems—*hypnosis*.

At the time, many investigators regarded hypnotic regression as a legitimate means of recovering suppressed memories. Later, cognitive psychology would show that hypnosis increases confidence in memories without improving their accuracy. The *Hills* weren't lying, but this makes literal interpretation far more complicated.

The *Hill* case remains foundational because it changed the tone of the phenomenon. Contact became known as clinical, involuntary, and frightening.

Whitley Strieber

By the time *Whitley Strieber* published *Communion* in 1987, the abduction narrative had grown darker and more psychologically complex. Earlier cases, such as the *Hills*, had already established the framework; *Strieber* transformed it into something far more ambiguous.

He described recurring nighttime visits from what he called "the visitors." These experiences involved bedroom intrusions, missing time, non-verbal communication, and medical-style procedures. The entities were clearly non-human, watching, intervening, and disrupting his life.

What made *Strieber's* story so impactful was his openness about uncertainty. He honestly considered explanations like sleep paralysis, hypnagogia, altered states of consciousness, and neurological anomalies. He never insisted the visitors were aliens and was transparent about his fears, doubts, and confusion.

This honesty gave *Communion* credibility. The famous gray face on the cover likely changed how people imagine alien beings more than any government report ever did. *Strieber* became an intermediary figure, neither fully believing nor fully skeptical, standing at the threshold between psychological explanation and ontological mystery.

Kelly Cahill

In 1993, in Victoria, Australia, *Kelly Cahill* and her husband reported one of the strongest multi-witness abduction-adjacent cases in modern ufology. Unlike many famous abduction stories, the case did not depend entirely on hypnosis, which gave it unusual evidential weight.

They shared their experience of seeing a large unidentified flying object floating over the road, with smaller craft nearby and tall, black-clad humanoid figures. *Kelly* later mentioned missing time, nightmares, and feeling very anxious after the encounter. Other nearby witnesses also saw strange lights that same night, which helps support their story, something that isn't always common in similar reports.

The real challenge is figuring out what's a stimulus and what's an interpretation. Witnesses might see lights in the sky, but that doesn't mean they've seen aliens. Just because something unusual happens doesn't prove it's from another world.

Researchers have different opinions. Some believe this is a strong example of a real anomaly, while others think stress, expectations, and memory reconstruction can explain it. This case highlights a common challenge in contact research: confirming that an event that's happened is not the same as understanding what it means.

The Allagash Abductions

The *Allagash* case from 1976 is often cited as one of the most compelling group-abduction reports. Four men on a camping trip in Maine observed a bright light hovering over a lake. They later realized they had experienced *missing time.*

During hypnotic regression, all four described being taken aboard a craft, encountering grey-type beings, and undergoing medical examinations. The consistency of their accounts made the case especially compelling to believers. Multiple witnesses telling the same story appear stronger than a single isolated report.

However, critics contend that group consistency may arise from common discussion, suggestions, and contaminated memories. Hypnosis remains a key factor: when one individual introduces an image or detail, others might unconsciously adopt it, particularly during memory reconstruction.

The *Allagash* case is key in debates about hypnotic regression. Does consistency show real memory, or just social influence? The case is still unresolved, caught between strong evidence and weak methods.

David Huggins

David Huggins represents perhaps the most psychologically complex category of contact experience. He claims lifelong interaction with non-human entities beginning in childhood, including sexual and emotional relationships, hybrid offspring, and affectionate, voluntary encounters.

Unlike many abductees, *Huggins* does not describe victimization in conventional terms. His accounts are intimate, symbolic, and deeply personal. He is especially known for explicit paintings that depict these encounters, transforming private experience into visual testimony.

There is little evidence to support his case, and it is rarely cited as proof of alien visits. Most often, experts analyze it as a psychological issue, involving trauma, imagination, dissociation, or symbolic stories. Some believe these encounters reflect unresolved psychological problems rather than real events.

His case raises an ethical question: when does belief become a pathology? Dismissing him risks ignoring genuine psychological experiences, but accepting literal claims without examination is also dangerous. *Huggins* is significant because he compels us to confront the line between belief, mental health, and meaning.

Comparative Analysis

Collectively, these cases demonstrate patterns and transformations. Certain themes persist throughout the years: humanoid figures, lost time, telepathic connections, altered consciousness, and the feeling of entering an unusual reality. Yet, the emotional tone differs significantly.

Angelucci's encounters are spiritual and uplifting. The *Hills* evoke fear and violation. *Strieber* introduces ambiguity. *Kelly Cahill* and *Allagash* raise questions of corroboration and memory. *Huggins* pushes the phenomenon inward, towards intimacy and symbolism.

This progression shows that contact experiences are tied to culture; they evolve with it. Early encounters focus on external skills and moral beliefs. Later ones increasingly involve bedrooms, dreams, paralysis, and consciousness. The experience becomes more internalized over time.

If the same visitors come over the decades, why do their cultural expectations change so much? Why do 1950s entities look like postwar spiritual optimism, while 1980s entities show existential fear and medical concerns?

The data suggest that human thinking mediates the process. Whatever the cause, whether an external anomaly, a brain event, or a psychological issue, the way it is understood is subjective and depends on existing mental symbols.

Competing Interpretations

Believers claim that consistency across cases indicates a real external phenomenon. Repeated themes suggest contact is genuine, even if not well understood. Variations are attributed to different species, motives, or consciousness, which serves as the medium of contact.

Skeptics see something else: the cultural evolution of a myth. They cite sleep disorders, hypnosis, memory reconstruction, and expectation effects. From this perspective, people are not lying; they're sincerely interpreting anomalous experiences through the strongest narratives available to them.

A growing middle-ground approach recognizes both the reality of the experience and the uncertainty about its cause. It doesn't insist that encounters involve literal extraterrestrials to take them seriously. Instead, neurology, trauma, dissociation, hypnagogia, and symbolic thinking can all create experiences that feel completely real to those who have them.

This may be the most analytically honest position, distinguishing between subjective authenticity and objective proof.

None of these cases provides falsifiable evidence of NHI. There are no recovered bodies, no independently verified medical evidence, and no artifact that definitively closes the question. Yet dismissing them outright overlooks their real significance.

These reports are based on personal experiences. They show how people make sense of the unknown and how culture shapes the interpretation. The fact that these stories appear across different times and places suggests they don't stem from a common external source but from a shared way of thinking.

Why do people across generations keep experiencing moments that feel like a special visit? Whether that feeling comes from the stars, the unconscious mind, or the space between them, it remains one of the great unresolved mysteries.

Chapter Seventeen

Speculative Technology

The Pyramid Network

What if the pyramids were part of a highly complex planetary system? It's interesting to think that a lost, non-human civilization may have built them to control Earth's environment and keep the Moon in its life-supporting orbit.

The Moon helps keep Earth's tilt stable, influences the tides that are vital for life, and slows down our planet's rotation to create days that are just right for living things. Getting the Moon positioned perfectly and safely would have been quite a challenge, and even small mistakes could have led to serious problems. In this story, we speculate that the pyramids are calibration points on the ground that link the Earth's crust, atmosphere, ionosphere, and our Moon.

The system could work in different ways. Pyramids might have acted as engines that interacted with Earth's atmosphere and ionosphere, influencing cloud formation, electric fields, and climate in small ways. They could also have sent signals or resonances to the Moon to correct minor errors, reduce tides, and prevent instabilities. The shape of the pyramid helps create standing waves in Earth's layers, turning large stone structures into planetary tuning devices.

The system would be entirely passive, with no engines, computers, or power plants. It would depend only on planetary physics, activating in extreme situations and then going dormant. Later, people saw only stone monuments, mistaking them for ritual sites or sacred designs, unaware of their true function.

This hidden system might still be detectable if you know where to look. Pyramids are located at mid-latitudes near *Schumann resonance peaks*. Schumann resonance peaks are natural electromagnetic "background hums" of Earth, created when lightning bounces energy between the ground and the upper atmosphere. The strongest peak is around 7.8 Hz, with others at higher frequencies.

Lunar mascons and small irregularities might be designed to reduce vibrations. Basically, the Pyramids align with Earth's natural electrical cycles, and the Moon's uneven features could be intentional adjustments.

The widespread presence of pyramids suggests a system driven by harmony and physics. Although surface erosion and myths can obscure their purpose, patterns of irregularity and resonance might offer helpful clues.

The idea is important: if humans were to build a planet stabilizer today, geometric shapes would naturally reemerge. Pyramids create a system that looks passive, scattered, and resistant to civilization. It suggests that Earth, and maybe the Moon, was purposefully designed.

Human Interface

Recent experimental work using *Synthetic Aperture Radar* (*SAR*) Doppler tomography has offered a provocative new perspective on this speculation. A team led by *Filippo Biondi* used satellite-based radar analysis of the Giza Plateau, processing electromagnetic signals reflected from the pyramids to produce high-resolution three-dimensional models of their internal and subsurface structures.

Their initial findings indicate the presence of previously unknown underground features, including chambers, vertical shafts, and pillar-like structures, that extend significantly deeper beneath the plateau.

The technique, which analyzes fine-scale vibrations and background seismic signals rather than direct excavations, treats the pyramid complex as a transparent object, thereby revealing density variations hidden within the stone and bedrock.

Although these results remain debated and require independent verification, they suggest that the pyramids are surface markers of a much larger underground system.

Within the speculative framework described here, these hidden structures are part of a larger planetary network linking surface features to deeper layers beneath the Earth. The visible pyramid is just the tip of a much larger structure beneath Earth's surface, supporting the idea of passive, planet-wide processes operating through resonance.

Some see the granite coffer as a body container, but also as a point of connection for a living person. In this chamber, human physiology links to a larger physical system. The occupant would serve as a biological marker within a structure that aligns with planetary forces, geometry, and perception, reflecting the pyramid's alignment with Orion. This chamber would connect human experience to Earth's movements and a fixed star pattern, using the sky as a reference point for orientation.

The dense stone, confined space, and precise proportions act like parts of a tool that affect electric fields, sound vibrations, and chemical states inside the body. Consciousness would be a measurable response of the nervous system within a carefully designed environment. The effect would be functional: a mechanical interaction among matter, energy, and perception that leads to changes in awareness in response to physical conditions.

The Holographic Universe

Modern physics suggests the universe might be even stranger than we think. According to the holographic principle, which is influenced by M-theory and string theory, all the information in a three-dimensional or higher-dimensional space could be stored on a lower-dimensional surface, like a hologram. Like a mathematical surface or network of information that gives rise to our 3D or 4D universe.

Hologram

In black hole physics, the event horizon acts like a boundary that holds all the information inside the black hole. In bigger cosmological ideas, the hologram could be part of the quantum structure of spacetime itself. Space, time, and gravity come from patterns of quantum information and entanglement.

This suggests that what we perceive as reality is an evolving phenomenon.

Analogies can really help make this abstract idea a bit more relatable. Think of a snow globe: the tiny 3D world inside is like our universe, and the globe's surface holds all the information needed to recreate it. Or consider a DVD: it's flat and 2D, but it contains the data to create a full 3D movie. In the same way, the universe's "code" is stored in an invisible network of quantum bits and relationships that shape everything we see and experience.

Is everything we see—stars, planets, even ourselves—a 3D projection generated by a hidden cosmic "code" beneath reality?

Picture your favourite video game. The 3D world on the screen, the trees, houses, and characters, feel so real as you play. But behind the scenes, all that action comes from data stored on a flat disc or in your computer's memory. The characters don't physically sit on the disc, but the data completely controls everything they do.

Some scientists think of our universe as a giant game. This invisible network or hidden code is the true blueprint of reality, and the world we experience is just what we interact with. In theory, everything in our universe could be read from this underlying code, like a game created from a disc.

From our point of view, everything happens within the 3D spacetime we experience: stars form, particles collide, and consciousness functions as usual. However, at a deeper level, all these events are stored in an underlying informational network, which makes the holographic layer more fundamental than the space we live in.

This way of looking at things has some interesting implications: if the universe is really made of information, then consciousness, remote viewing, and other unusual phenomena might affect reality through its informational structure. This idea helps connect physics, consciousness, and what many consider paranormal events.

Wireless Energy

Another interesting idea suggests that the Giza complex might actually be part of a wireless energy system. Even before modern power grids, *Nikola Tesla* believed that electrical energy could be sent wirelessly by resonating between the Earth and the ionosphere. In this perspective, the planet itself would act as a conductor, with specially tuned structures serving as points for transmission or connection.

The Great Pyramid's materials and shape really make you think. Its main structure is built from limestone, which has interesting electrical properties. Inside, there are chambers made of granite rich in quartz, a mineral that can generate an electric charge when squeezed or stressed. So, if there are vibrations from the environment, earthquakes, or electrical changes in the air, these materials could, in theory, respond to electric fields in an organized way.

Wireless Technology

Within this speculative idea, the pyramid's shape might be seen as a big resonator, built to reflect or change naturally occurring electrical potentials between Earth and sky. The downward and upward passages could act like channels or waveguides, while the central chambers might serve as spots where energy is gathered or redistributed.

The tall, pointed forms of obelisks make them practical components of this system. Shiny granite or diorite surfaces might act as vertical conductors, capacitive antennas, or resonance amplifiers, channelling electrical energy between the ground and the sky.

Obelisks are carefully placed and sized to fine-tune the planetary network, boost standing waves, or focus energy at key locations throughout the area. They can also serve as helpful reference points, aligning energy flow with celestial patterns such as Orion's Belt to ensure the network operates smoothly within a clear space and cosmic framework.

Unlike modern power plants, this system wouldn't need traditional fuel. Instead, it would use natural planetary energy, such as telluric currents, atmospheric electricity, and electromagnetic waves between the ground and the ionosphere. Energy wouldn't be made but *collected*, focused, and shaped by the structure itself.

From this perspective, the pyramids and obelisks function as passive transformers or components of a distributed, planetary-scale energy system. While it remains unproven whether such a network could transmit usable power, it demonstrates how monumental construction might be envisioned as part of an electrical infrastructure that interacts with the planet's natural fields.

This interpretation relies on known physical phenomena, resonance, conductivity, and electromagnetic coupling, applied on a large scale, without involving exotic forces or supernatural ideas.

Vajra and Plasma Technology

Another aspect of this fascinating framework is plasma, the electrically conductive state of matter that exists in both space and our atmosphere. The ionosphere is a layer of plasma shaped by Earth's magnetic field and sunlight, and it naturally supports standing electromagnetic waves.

If huge structures like pyramids and obelisks were meant to work with atmospheric electricity, plasma dynamics might have been part of their function.

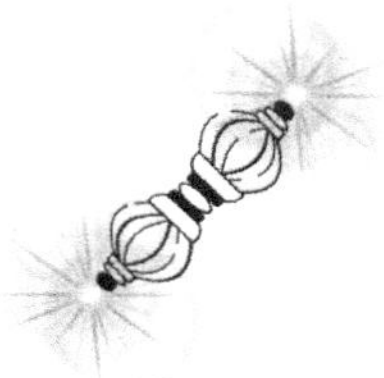

Plasma Technology

Ancient symbolic traditions might preserve traces of this understanding. In Indian cosmology, the Vajra, often described as indestructible, luminous, and thunderous, has traditionally been regarded as a divine weapon.

However, its symmetrical, axis-like shape and connection to lightning suggest a different way of looking at it: not just as something destructive, but as a symbol of controlled energy release and stability. Lightning is powerful and helps balance the electric potential between the sky and the ground.

In this view, the Vajra symbolizes the control and direction of plasma, using a structure to contain energy rather than letting it flow chaotically. The repeated appearance of similar symbols across cultures might indicate an instinctive understanding of plasma as a force connecting Earth and the atmosphere, one that can be shaped by geometry, alignment, and materials rather than machines.

Jody Swannell

Feedback

If the world were designed by humans, it wouldn't be static; it would need feedback to stay balanced. In modern engineering, complex systems always adjust themselves. Thermostats control temperature, gyroscopes keep spacecraft stable, and power grids quickly adapt to changes in demand. Stability is continuous, maintained through ongoing adjustments.

Earth naturally keeps its temperature in check through various systems. The oceans play a big role by absorbing and redistributing heat. The carbon cycle helps manage greenhouse gases, while plate tectonics contributes to climate stability by recycling carbon into the Earth and releasing it through volcanoes. These automatic processes work behind the scenes to prevent extreme changes and help create conditions that are just right for life.

The Moon helps keep Earth's tilt steady, preventing big shifts that could cause extreme climate changes. Its gravitational pull affects ocean mixing, nutrient movement, and possibly early life. The orbit of the Moon around Earth creates a balanced system that avoids dangerous drift.

In simple terms, this is like a group of connected oscillators: objects that influence each other by exchanging energy and adjusting their movements through resonance.

In this chapter's speculative view, large structures such as pyramids could serve as calibration points, built to align with Earth's natural standing waves. They would gently interact with atmospheric or lithospheric oscillations. If such a network existed, it would help balance things, activating only when thresholds are crossed and returning to a resting state once balance is achieved.

The system's intelligence depends on how it's designed. Stability occurs when its shapes and planetary physics work together. The key idea is feedback: a rational world must sense what's happening, respond, and fix itself.

Time Calibration

Any engineered system must also account for time.

Earth isn't just spinning and orbiting; it's also going through a bunch of interesting processes. Over about 26,000 years, its tilt gradually shifts, a phenomenon called *axial precession.* The shape of its orbit changes slightly, and the Sun's output varies too. The distance between Earth and the Moon gradually increases as the Moon moves farther away. Across thousands of years, climate patterns also shift in predictable rhythms.

Ancient monumental architecture often encodes astronomical alignments such as solstices, equinoxes, stellar risings, and precessional markers. Although often viewed as symbolic or ritual, such structures may also serve as long-term calibration references. A fixed stone structure aligned with celestial events is a clock linked to cosmic cycles.

In engineering, this is called *phase-locking*, which means matching a system to an external oscillator.

When a repeating system synchronizes with a more stable external rhythm, the two cycles interact, and one gradually adjusts until they move in harmony, maintaining alignment over extended periods. Instead of drifting apart, the systems stay coordinated through resonance and timing.

If Earth were part of a larger engineered system, celestial alignments would create stable reference points for understanding terrestrial processes. The sky would act as a coordinate system, with Orion, the ecliptic, and the polar stars serving as calibration markers.

Unlike digital memory, stone endures for millennia. Geometry resists erosion, and a structure aligned with astronomical cycles becomes a time-storage device, a physical record of the phase relationships between Earth and the cosmos.

The pyramids and similar monuments could link Earth's natural processes to the regular cycles of the cosmos. Time acts as a kind of blueprint in engineering.

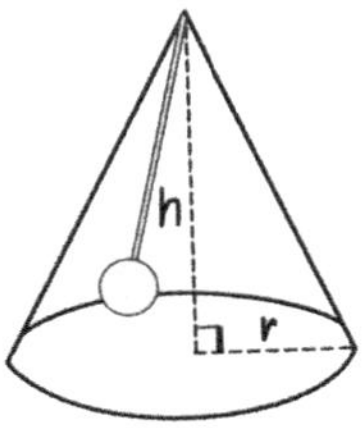

Pendulum

Sometimes, natural alignments lead to small, noticeable anomalies. During several total solar eclipses, some paraconical pendulums, delicate devices used to detect tiny changes in Earth's rotation, exhibited subtle deviations from their expected motion. In a few cases, the pendulum's swing seemed to drift in the opposite direction, as if local inertial conditions had briefly reversed.

Although debated and not fully understood, these observations suggest that accurate Sun–Moon–Earth alignment may affect terrestrial systems in ways that current models don't fully account for.

If even a fraction of these effects is genuine, they suggest that celestial geometry does more than mark time symbolically. Alignment might subtly but measurably influence gravitational gradients, atmospheric loading, electromagnetic conditions, or inertial frames.

A system designed to keep planets stable would see eclipses as stress tests, moments when alignment puts pressure along one direction, which could cause brief fluctuations.

Eclipse

In this context, astronomical calibration serves as a reference point. Rare alignments serve as helpful checkpoints in the system, points where we can confirm the phase across celestial bodies. This shows how closely Earth is connected to the larger gravitational and electromagnetic environment.

Time, therefore, encompasses brief intervals of compression during which the geometry temporarily amplifies interactions between planetary layers and cosmic bodies.

Part Six

CONVERGENCE

Chapter Eighteen

Remote Sight & NHI

Remote viewing began as a way to test whether perception could extend beyond the physical body. Across many sessions, some descriptions matched their targets more often than chance would expect. In this context, the target was seen as a fixed, separate entity. But during a few sessions, something unexpected happened.

A few viewers noticed a difference in the overall experience. Instead of feeling like they were just watching something distant and passive, they felt a presence. The target didn't seem like a mere place waiting to be described anymore; it felt aware. A few participants even shared an odd feeling that their act of observation was recognized.

These accounts came from different individuals and experiments. They were detailed carefully and often with caution. However, they suggested that perception could involve multiple aspects.

In the standard model, the observer just perceives things, and the target exists. Information flows in one direction. But when attention notices something that responds even a little, the exchange becomes more than just observation. It turns into a kind of interaction, and the line between observer and observed starts to blur.

This is where convergence begins to take shape. Distance matters less than alignment. Contact doesn't require physical proximity; it occurs when awareness and design align in the same pattern. Under certain conditions, perception may involve reciprocity.

Starting in the late twentieth century, people from different countries began sharing stories of encounters that had many similar features. Common elements included paralysis, glowing surroundings, nonhuman figures, medical-style procedures, telepathic communication, and missing time. Many described these experiences as remarkably real, physical, and emotional.

Researchers like *John E. Mack* approached these accounts cautiously, documenting similarities without assuming they were caused by pathology or external visitors. What he found was that many experiencers showed no signs of psychosis, and their lives continued normally. Nonetheless, the encounters left a lasting psychological impact.

The phenomenon existed in a bizarre realm. It defied classification as a dream, resisted reduction to fantasy, and challenged conventional physical explanations. Abduction reports appeared less as isolated hallucinations and more as a widespread shared experience.

These accounts show an important difference. Remote viewing suggests that consciousness can expand, while abduction narratives suggest that contact can intrude.

Whether considered as outside intelligence, altered states, archetypal activation, or interactions with internal consciousness, the main features remain the same: body paralysis, heightened awareness, a sense of presence, and telepathy.

Abductees

Contact through resonance within consciousness is taken seriously; case studies show how such alignment might be experienced. Over many years and across different parts of the world, people who don't know each other have shared encounters that are described as surprisingly similar, even when the details differ.

In 1975, logger *Travis Walton* went missing for five days after reportedly seeing a bright craft in Arizona, an observation made by several coworkers. *Walton* later recalled waking up in a luminous environment, interacting with nonhuman beings, and experiencing a state that seemed to lie between physical reality and an altered state of consciousness.

The case attracted considerable attention because multiple witnesses confirmed the initial sighting. The incident was later popularized through the 1993 film *"Fire in the Sky,"* which dramatized *Walton's* story and increased public interest in the case. Notably, the men who were with him that night consistently recalled the initial encounter, reinforcing the idea that what happened was a shared event that later diverged into *Walton's* individual experience.

In 1989, *Linda Napolitano* claimed she was taken from her high-rise apartment in New York City, in front of witnesses who reported seeing a glowing object outside the building. Her account included paralysis, levitation, and structured beings communicating telepathically. Investigators debated the physical aspects of the case, but the experiential elements were consistent with earlier reports.

In China, *Meng Zhaoguo* recounted a 1990s encounter involving luminous objects, physical effects, and interactions with a nonhuman female entity. His story drew national attention and was investigated by local researchers. The account combined physical details with a deeply personal transformation, highlighting the widespread nature of similar encounter themes worldwide.

In 2001, Australian experiencer *Amy Rylance* reported vanishing from her locked bedroom in rural Queensland and later described time aboard a craft with tall, luminous beings. Those present in the home at the time claimed she physically disappeared before reappearing days later. As in other cases, the core elements included altered awareness, entities, and a sense of purposeful interaction.

Former president of the *International Chess Federation, Kirsan Ilyumzhinov,* publicly stated that he was taken aboard a spacecraft from his Moscow apartment in 1997. He described telepathic communication with humanoid beings dressed in yellow suits and framed the event as calm and instructional rather than threatening. His professional status added an unusual political dimension to the phenomenon.

Even actress *Fran Drescher* has shared her experience of an abduction she believes occurred during her youth, describing a period of missing time and a scar she links to the incident. Her account, like others, combines personal memory, physical signs, and firm conviction.

Abduction

These individuals come from very different backgrounds, cultures, and beliefs. Some were eager to seek publicity, while others preferred to stay out of the spotlight. Some stories mention supposed witnesses, while others happen entirely in private settings. Despite these differences, certain elements keep recurring: paralysis, glowing environments, strange nonhuman figures, telepathic communication, time loss, and lasting psychological effects.

If contact relies on spacecraft travelling across interstellar distances, then these accounts require immense logistical effort. Maybe no one actually goes anywhere. Perhaps the body stays in place, and the experience occurs because the mind enters a state in which it perceives another presence. Instead of travelling through space, it's *contact through awareness*.

Systemization

In this scenario, the phenomenon depends on consciousness's capacity to align with intelligence. The convergence happens where awareness and design intersect, rendering distance irrelevant, as the encounter becomes a matter of experience rather than mechanics.

If these events are seen as converging rather than travelling, many otherwise puzzling aspects become clearer.

First: Paralysis

Across cultures, experiencers report an inability to move while awareness remains heightened. Neurologically, this resembles sleep paralysis, a state in which REM atonia persists even as consciousness is maintained. The body remains inhibited as awareness shifts into an alternative perceptual configuration. Immobility may serve as a protective mechanism, preventing physical interference while cognition engages elsewhere.

Second: Telepathic Communication

In these encounters, words are rarely spoken. Instead, information comes in all at once: ideas, feelings, and images. This is similar to how insights appear in dreams or deep intuition: quick, layered, and multidimensional. If the connection is based on resonance, language might not be the best way. Instead, alignment could transfer information directly, without using words.

Third: The Environment

Many describe bright interiors without visible lights, featuring curved shapes, smooth surfaces, and a sense of spatial distortion. These environments seem built, but not in a mechanical way. If awareness connects with sensible design, "space" might be an interface.

Remote viewing proposes that consciousness can reach beyond the physical body. Abduction reports indicate that an external entity may interact with it there. Both concepts suggest that awareness extends beyond the skull into a wider information field. Under specific conditions, such as trauma, altered states, focused or auto-triggering, this field can become interactive.

Consistency among experiencers is crucial here. The beings differ in appearance, small and grey, tall and luminous, insect-like, human-like, yet the central elements stay the same: paralysis, examination, communication, and return. This suggests that the surface imagery might be open to interpretation, while the encounter remains the same.

If convergence is real, it might not happen entirely by choice. Just as a radio can both transmit and receive signals, consciousness may also be able to project itself and be accessed. Usually, the system remains silent, but sometimes alignment occurs.

Importantly, this model includes psychological explanations. The mind isn't creating the event, but it shapes how we experience it. The encounter isn't just outside or inside us, but something that happens between a person and their experience.

This explains why physical evidence is hard to find. Since the main exchange happens through consciousness, physical signs would be small, inconsistent, or absent. The observable world might only show indirect effects, such as skin impressions, electromagnetic changes, and bodily stress.

The convergence hypothesis is somewhat of an awkward fit. It doesn't quite align with materialist ideas, nor does it require spacecraft to travel across galaxies. Instead, it proposes that, if contact occurs, it draws on abilities already present in human cognition.

A change in consciousness can facilitate encounters with NHI via remote sight, removing the need for interstellar travel. Scientific understanding must recognize that consciousness itself is still not fully understood.

Perception

Neuroscience views the brain as a predictive organ that continually constructs and updates models of reality based on incoming information.

Normally, sensory input keeps these models in check. However, they aren't as strict during altered states like trauma, sleep paralysis, deep meditation, or psychedelic experiences. Consequently, the line between imagined images and external information becomes fuzzy. What seems to be outside may actually come from the interaction of expectation, memory, and unusual stimuli.

Research on sleep paralysis helps illustrate how the brain behaves during sleep. During REM sleep, the body is neurologically shut down, yet vivid images and sensations persist. When someone wakes during this phase, they often feel paralyzed, sense a presence, experience chest pressure, and see glowing figures. The sensation of an intruder is common across many cultures and aligns with findings from brain research. However, neuroscience doesn't fully explain why this sensed presence seems to act on its own, as if it has purpose and awareness.

When the brain encounters ambiguous or unfamiliar input, it may generate an agent as the most efficient explanation. Agency detection is deeply embedded in human cognition; we are wired to assume intention behind patterns. Under high uncertainty, the mind might default to perceiving an intelligent presence.

Agent

Field-based theories of consciousness add another layer. Some researchers suggest that consciousness is not generated solely by the brain but involves interaction with broader electromagnetic or informational fields. From this perspective, the brain functions as a *two-way transducer*, receiving input from these fields, filtering and tuning it, then sending coherent signals back.

Although it's still a topic of debate, this model offers a helpful way to understand phenomena such as shared odd experiences, telepathy-like sensations, and reports of remote perception, implying that the mind might both sense and influence information beyond the body.

Even ideas from quantum cognition, which are often misunderstood in popular science, can help us understand these experiences. At very small scales, systems can exist in multiple states at once and only settle into a specific state when they interact with something. Similarly, our mental states can hold multiple possibilities before one becomes clear. An encounter might just be a temporary alignment between minds, a kind of synchronization, rather than a literal journey from one place to another.

None of these ideas confirms abduction stories. But they expand the range of ideas in which such experiences can occur. They show that perception is flexible, context-dependent, and state-dependent. They suggest that awareness is more adaptable than classical materialism suggests.

In this broader view, convergence is more about an arrangement than something mysterious. An encounter happens when the mind adjusts and connects with another presence. The experience feels real because the mind and body react, even if we don't fully understand it.

Remote-viewing experiments indicate that discernment can separate from direct sensory input. Abduction stories propose that this separation might become mutual. Neuroscience demonstrates that the brain can create immersive, agent-filled realities when conditions are altered.

Field theories propose that thought and awareness may extend beyond the brain itself. The convergence hypothesis combines three concepts: consciousness as a medium, perception as interaction, and contact as a meeting within a shared informational field.

It appears that the human mind can genuinely connect with a presence beyond itself, even when that presence exists outside the usual laws of matter.

Chapter Nineteen

Divine Encounters

The Sanskrit word *avatar* (avatāra) literally means "descent." In Hindu tradition, it denotes the physical manifestation of a deity or powerful spirit in the material world, which may or may not take human form. Earlier, avatars were considered divine beings who manifested as mysterious lifeforms, sometimes indigenous ones. Now, we see an avatar more as an interface influenced by its environment rather than as the divine itself.

This recurring avatar motif surprisingly fits with simulation and holographic universe models without requiring belief in gods as traditionally understood. In these models, higher-level systems don't completely appear in lower layers of reality. Interaction occurs through partial representations that adhere to local physics, biology, and limits of perception.

What ancient cultures called gods might have been encounters with intelligences that only interacted under certain conditions, rather than all-powerful beings manipulating the world.

Avatar

Most of the earliest myths depict gods taking on different personifications. Vishnu, for example, doesn't appear as a single human form but in various embodiments, including fish, turtle, boar, and human, each suited to a specific challenge.

Many traditions depict divine beings as animals, burning bushes, or pillars of light. Sky figures occasionally descend to warn or impart knowledge, then vanish again. Watchers observe humanity from afar but rarely rule directly or take physical form. This pattern suggests restraint rather than freedom, as if these beings can appear to humans only in uninhibited ways.

In a simulated reality, this behaviour makes sense. Higher intelligence cannot fully express itself without disrupting the lower levels. Interaction requires simplified images that align with biological and mental limitations. Avatars aren't disguises but appropriate representations.

A fish in a flood narrative functions more pragmatically than symbolically: aquatic life endures when humans cannot; a fish can move easily through a collapsing environment; and its presence remains understandable within mythic logic. The avatar symbolizes continuity when systems fail.

This shows why it's better not to have too much direct contact all at once. If reality has many layers, being fully present might overwhelm us and confuse its meaning. Trying to make everything clear too quickly could upset the balance.

So, it's best to step in only when needed and understanding is possible.

Ancient myths often use cautionary tales, emphasizing guidance over control and focusing on saving small parts rather than entire civilizations. These stories show that an intelligent force isn't rescuing us but that it is willing to step in to help.

In holographic models, after a civilization falls apart, technical knowledge fades, but symbols remain, and interfaces become deities. The *same entities* appear across cultures in different forms. Myths are symbolic records of encounters that are now misunderstood.

Reset

By "*reset*," I mean times when civilization faces big moments of change, when old systems of knowledge fall apart, and we must rebuild. Throughout history and in stories, these events include floods, plagues, ice ages, large migrations, the fall of empires, and periods of technological slowdown.

During hard times, writing can disappear, institutions can fall apart, and all the knowledge we've gained can be lost. Astronomy might become astrology, and engineering might become sacred geometry. Practical skills often fade, replaced by symbols and memories. What remains are stories that people can recall when official records are gone.

The phrase "before and after resets" explains how extraordinary or transcendent experiences stay with us. Before the collapse, these encounters, whether real, imagined, or misunderstood, could be viewed in organizational or observational terms. The figures, angels, gods, avatars, and *Watchers* remain recognizable, but their meanings change as our understanding of them evolves.

What stays the same over time often shows how we naturally respond to things we don't understand. It's a familiar part of life that shapes similar experiences and reveals how our minds deal with things we can't fully comprehend. Continuity acts as a bridge, linking what we know with what remains a mystery as history is forgotten and starts over.

Avatars stay the same across cycles even when the interface changes. This is similar to unexplained flying objects: they observe, intervene at key moments, and don't escalate.

Book of Enoch

A darker version of this pattern appears in the ancient Jewish text known as the *Book of Enoch*. Written roughly between 300 BCE and 100 CE, it describes a group of angels called *Watchers* who descend to Earth and violate their mandate by directly engaging with humanity. Two hundred *Watchers*, led by *Semjaza* and joined by *Azazel*, swear an oath and take human wives.

Their offspring, the *Nephilim*, become violent giants who consume resources, devastate the land, and resort to cannibalism. The *Watchers* also impart forbidden knowledge, including metallurgy, warfare, sorcery, and celestial secrets. This knowledge accelerates humanity's collapse.

Divine judgment follows the archangels' report of the corruption to God, and the *Watchers* are bound and imprisoned. The *Nephilim* destroy one another, and their spirits become wandering demons. Humanity endures, but only through catastrophe. *Watchers* don't rely on restricted interfaces; they're seamlessly embedded in human life.

Enoch serves as the ideal corrective model. Chosen as a righteous intermediary, he is charged with warning the fallen angels and is guided on visionary journeys through the cosmic realms. Angels reveal to him the movements of the sun and moon, the laws governing the seasons, and the structure of reward and punishment.

At the conclusion, *Enoch* becomes a radiant heavenly being, serving as an authorized intermediary between humans and the cosmic realms. Unlike the *Watchers*, who failed to embody their roles, *Enoch* succeeded by mediation.

The *Book of Enoch* shares an ancient story about serious troubles. It explains that human suffering isn't just due to human mistakes but also to interference from non-human beings. Having knowledge without being ready can cause harm. The story advises caution with contact and hybridization, as it can upset both worlds.

Across different cultures, divine encounters tend to follow similar patterns. Supernatural beings are rarely present; communication is symbolic; knowledge is controlled; and visitors leave after their intervention. Whether called avatars, angels, or *Watchers*, the idea is the same: higher beings, although bound, reach out to restore harmony to humanity.

Ancient gods are like quick snapshots of a simpler part of reality, showing us an important truth: the patterns we see reveal that people interpret the same events differently over time. What really advances is our ability to see the intelligence behind these experiences. Recognizing this helps us move beyond superstition toward a deeper appreciation of *universal consciousness*.

Ethiopian Bible discoveries

The *Ethiopian Orthodox Tewahedo Church* has one of Christianity's most extensive and ancient biblical canons. It preserves the complete text of the *Book of Enoch*, written in the classical Ethiopian language of Ge'ez long after it disappeared from Western scriptural traditions. Although sections of *Enoch* were later found among the Dead Sea Scrolls in the twentieth century, confirming its deep roots in Second Temple Judaism, Ethiopia has preserved the full narrative for centuries.

The Book of Enoch was far more than a fringe curiosity; it was a widely read text in early Jewish and Christian communities. Its detailed cosmology features multiple heavens, complex angelic hierarchies, and the descent of the *Watchers*, who cross an ontological boundary to engage directly with humanity.

Ethiopian tradition also preserves a powerful symbol of contact: the belief that the *Ark of the Covenant* is kept in Axum, guarded within the *Church of Our Lady Mary of Zion*. Whether taken as historical fact or as symbolic, the *Ark* in biblical stories represents a divine presence that demands strict protocols, as mishandling it can be fatal.

The stories emphasize a common theme across ancient systems: sacred technology can only be accessed if specific conditions are met.

Ethiopia was largely outside the control of the later Roman Church, so it did not follow the same process of removing certain books from the Bible. As a result, older religious ideas persisted there, whereas they disappeared in much of Europe.

Texts from other sources are still regarded as part of Ethiopia's scripture, offering a more detailed and complex view of divine encounters. In this tradition, meeting higher beings is more than a metaphor. When certain conditions are met, these encounters follow a specific order, but they are dangerous if not handled carefully.

Ethiopia's preservation of the Book of Enoch mirrors a common pattern seen across various cultures, such as Egypt, Sumer, and Mayan traditions. In these societies, interactions between different realms are restricted, knowledge is tightly guarded, and unauthorized frontiers often have negative consequences.

Gods of Egypt

In ancient Egypt, gods were seldom portrayed as fully human rulers walking openly among people. Instead, they appeared in powerful multiple forms, with human bodies combined with animal heads, luminous disks, or pure light, symbolizing their divine mystery and authority. These forms reflect the deeper restrictions and reverence associated with their presence.

Ra embodies the sun itself, travelling across the sky in a barque, entering the underworld each night and emerging reborn. His daily cycle symbolizes cosmic renewal. He wouldn't casually intervene in human affairs, as his presence could be catastrophic. Contact was established through priests, temples, ritual geometry, and symbolic embodiment.

Ra, Horus, Anubus, Thoth

Horus is represented as a falcon-headed figure, symbolizing vision. *Anubis* has the head of a jackal, connected to death's thresholds. *Thoth* appears as an ibis or baboon, animals associated with writing and measurement. Each form has a specific purpose and relates to a boundary within its domain.

The Egyptian system suggests that divine intelligence is expressed through symbolic simplification. A god isn't the full scope of intelligence; instead, it's a limited projection designed for interaction within specific ecological and psychological contexts.

Pharaoh himself becomes an avatar, a living vessel through which divine authority is expressed within the political organization. The king is not the source of divinity but its intermediary.

More central than any individual god is *Ma'at*, who exists both as a goddess and as the principle of cosmic order. She is often portrayed as a woman with an ostrich feather on her head, making the abstract idea of balance visible and tangible. Ma'at is not merely a moral concept; she embodies the structural order that enables the universe to function.

Ma'at

The rising sun, the Nile's flooding, the pharaoh's legitimacy, and even truth in human speech all belong to Ma'at. The gods themselves serve as custodians of this equilibrium. Their interventions are measured and precise. In this way, Ma'at is both the visible goddess and the unseen law of balance that maintains the cosmos.

This fits well with the layered-reality idea: high-order intelligence only shows up when it functions smoothly with the system.

Anunnaki gods

In Mesopotamia, the Anunnaki gods are described as descending from the skies to establish civilization. They introduced agriculture, writing, metallurgy, and law.

Enki generally acts as a mediator. He gives early warnings about the upcoming flood, protects important knowledge, and sometimes bends the rules a bit to keep things peaceful. On the other hand, *Enlil* is more about authority and rules, often suggesting destruction when humanity gets out of line.

The flood story, also echoed later in the Epic of Gilgamesh, illustrates divine intervention at a pivotal moment in the history of civilization. Humanity was noisy and chaotic, disrupting the cosmic order. One individual is warned that a reset is imminent.

Anunnaki

Sumerian deities were deeply involved in human affairs, unlike the Egyptian gods, who appeared distant. They shaped cities, designated kings, and granted the *me*—divine laws and powers that endowed humans with the knowledge and authority to lead civilization. Each one specialized in a specific field of society, such as law, leadership, crafts, and writing, emphasizing their vital role in shaping human progress.

The gods shared these gifts cautiously, as misuse could cause disaster. In Sumerian stories, divine intervention is direct and practical, offering tools for survival while sustaining order.

When higher-order intelligence shares too much technical knowledge prematurely, it can upset human society. In Sumerian stories, the flood acts like a reset button, washing away what's out of balance and making room for a fresh start.

Afterward, the gods step back, leaving behind the institutions they built, so civilization can continue.

Mayan Cycles of Creation

Mesoamerican cosmology gives a clear view of resets. The Maya believe there were several attempts to create humans before the current one. Each failed in a different way: one dissolved in water, another turned into chaos. After many attempts, humans were finally made from maize, demonstrating resilience and the enduring power of creation.

In the K'iche' Maya text Popol Vuh, the gods create beings who see everything clearly. Their perception is so sharp that the creators hide their vision to prevent over-understanding. This detail is important. Knowledge must be controlled because knowing too much can upset harmony.

The Maya gods are not entirely all-powerful as in a monotheistic view. They experiment, adjust, erase, and continually refine their actions.

The feathered serpent deity Kukulkan, also called Quetzalcoatl in central Mexico, is an intelligent entity associated with astronomy and cyclical time. He appears, imparts knowledge, withdraws, and is prophesied to come back.

Interventions

Across Egyptian, Sumerian, and Mayan systems, several constants emerge:

1. Divine beings rarely remain permanently in bodily form
2. Knowledge transfer is selective and potentially risky

3. Civilizations undergo resets due to an imbalance
4. Vision is intentionally obscured
5. Gods withdraw after their intervention

From a layered-reality perspective, gods aren't human-like rulers but simpler versions in which higher systems discreetly interact with people. Together, they describe a feedback loop between intelligence and civilization.

What ancient people thought of as gods might actually have been repeated visits from higher intelligences, remembered through stories after each major collapse. While the way they appear changes over time, the underlying motive stays the same.

The unnerving implication that different civilizations around the world developed the same rules and gods. Either humans naturally imagine the same patterns or the encounters themselves follow the same rules everywhere.

The presence of these beings can unravel cities or entire worlds in an instant.

Chapter Twenty

The End of Secrecy

Secrecy implies oporoprietorship. It assumes that someone, somewhere, possesses the complete truth, as if it can be stored, protected, and eventually disclosed. Hidden programs, classified briefings, compartmentalization, and the belief that information remains behind sealed doors, awaiting political approval for release.

Secrecy relies on restricting information, controlling events, and gathering the unknown into institutions to manage it. When the subject surpasses these boundaries, containment fails.

Governments may have information, such as unusual radar readings, recovered materials, and pilot testimony. They may also understand aspects of advanced propulsion or exotic physics that civilians can't access, but they don't own the phenomenon itself.

Across cultures and eras, witnesses report brief encounters, revelations, and strange experiences. The insights gained from these reports are sporadic.

Countries control territory and information flow, but

awareness develops within individuals. Inner experience surpasses administrative oversight. Researchers, physicists, experiencers, theologians, and military personnel are converging on similar conclusions from different viewpoints.

The end of secrecy has no defining moment. Pressure slowly mounts across different disciplines until the conversation expands. Physics questions the solidity of matter, while neuroscience views awareness with increasing humility.

Cosmology broadens the scope of what is possible. Separate paths of investigation begin to intersect, making convergence harder to dismiss as mere coincidence.

Intelligence is spread out, and consciousness starts to look like a fundamental part of reality. Humanity's sense of isolation decreases as patterns appear in myths, religion, strange experiences, and modern observations.

Humans share the same way of thinking, not a different world. Each generation uses its own symbols to understand itself, but the connection remains beneath the surface.

* * *

The end of secrecy means we all become more aware. As the veil lifts, we begin to see ourselves as part of a larger community of shared intelligence.

The excitement comes from growth and realizing that reality has many layers of awareness connected to ours as it unfolds. As our perspective expands, secrecy dissolves, unveiling new knowledge and exciting potential.

A common mistake we make when examining ancient religion is assuming that ancient people aimed to invent gods. We often read mythology as fiction because we inherit it as literature, but for those who lived in those times, these stories were frequently seen as memories, encounters, and warnings vital for their survival.

* * *

Throughout this book, we see a repetitive cycle: direct contact with higher intelligence is rarely depicted as unplanned. Instead, it's organized, controlled, and often perilous. Whether examining the *Ark of the Covenant*, the Egyptian priesthood, Greece's mystery schools, the *Watchers* in *Enoch*, or recent reports of UAP near nuclear sites, the same hypothesis appears.

We assume that having more information is always advantageous, that knowledge should be instantly available, and that access alone signifies progress. Ancient systems were much more guarded because they viewed knowledge as a potent force that could hurt individuals and societies if shared without expert preparation. Their course of action was managing risk. They recognized that wisdom and information are fundamentally different concepts.

This is especially important when considering how belief shapes reality. Throughout history, civilizations have understood that controlling knowledge is a form of authority. Myths were more than spiritual stories; they guided people's views on authority, threats, and the mysterious. By controlling storylines about reality, you can influence how people perceive it.

* * *

Governments, intelligence agencies, and private organizations all pursue interests in consciousness, remote viewing, psychological influence, and ways to influence public awareness. The logic is straightforward: consciousness isn't apart from power. When attention, beliefs, and expectations affect behaviour on a broad scale, controlling information becomes as crucial as controlling territory. The battlefield extends beyond the physical realm; it's also *cognitive*.

That's why the UFO /UAP question is so closely tied to religion, psychology, and myth. The same sight can be called an angel, a demon, an alien ship, a divine sign, or a secret military project, depending on the observer's perspective.

The object itself isn't as important as the meaning we assign to it. That's also why this phenomenon is so tricky to pin down: its existence is inseparable from people's ability to comprehend it.

* * *

Ancient cultures seemed to understand this intuitively. Contrary to separating the spiritual from the political, or consciousness from cosmology, they correctly assumed that reality included both visible and invisible systems of influence.

We modern people often divide these categories too sharply, and in doing so, we may miss the larger picture. A civilization can unravel when people no longer trust institutions, share meaning, or agree on what's real.

Myths outlast empires because symbols endure. Sacred stories can preserve warnings for thousands of years, unlike scientific reports.

This may explain why the same systems keep appearing in different civilizations. Ancient people copied one another, responding to the same kinds of encounters and preserved the information in the most persistent form available. Ritual, symbol, and story are stronger across time than institutions. A temple can fall, a library burn, but a myth survives in societal memory.

Humanity isn't isolated. Science fiction's idea of visitors

from another planet might be too naive; intelligence could exist on levels we barely sense. Consciousness could be part of and connected to reality. If so, divine encounters, mystical visions, and anomalous experiences don't belong in separate categories.

We are looking at different expressions of the same underlying relationship between human awareness and a larger field of intelligence.

Skepticism is necessary. But real skepticism means questioning both blind belief and blind dismissal. It means asking whether our modern assumptions are as insufficient as the ancient assumptions we criticize. We pride ourselves on being rational, but rationality without imagination can become its own form of blindness.

* * *

The ultimate purpose of this book is to show that some patterns are too persistent to ignore. Throughout history, across continents and belief systems, humans tell similar stories: modest contact, knowledge can be risky, consciousness is important, and imbalance leads to collapse. *That consistency deserves attention.*

The greatest veil is not secrecy but the assumption that ancient knowledge and modern inquiry inhabit separate worlds. They do not. They are part of the same human effort to understand who we are, where we are, and whether we are alone.

The reader doesn't need to accept every conclusion to feel the weight of the issue. We have called it divine, supernatural, extraterrestrial, symbolic, and psychological. The name matters less than the recognition that the phenomenon persists.

Civilizations separated by oceans wouldn't, by accident,

keep telling the same story. We shouldn't keep rediscovering the same warnings if there is nothing there to warn us about.

The final truth is humanity has always been in a relationship with something larger than itself, and every age has struggled to translate it.

Simple Answer

For ages, we've separated reality into categories like science and religion, matter and spirit, mind and machine, gods and aliens, myth and fact, often assuming truth belongs to one side. However, the "theory of everything" isn't just a single equation on a blackboard, but the understanding that all these pieces form one interconnected larger picture.

Physics tells us that matter is composed of patterns of energy connected by relationships. Consciousness can't be explained by chemistry alone. Ancient religions believe that intelligence exists beyond what we see, and mystical traditions think reality is shaped by awareness.

Some theories suggest that what we call physical life might just be an interface. Even though these ideas are different, they all suggest that reality is more like a mind than a machine.

If consciousness is fundamental, everything begins to look different. Mind isn't something the universe creates; it's woven into the fabric of reality. Observation is important because awareness is part of reality's structure. The old idea that the universe is alive no longer sounds ridiculous.

Gods, angels, Watchers, ancestors, and even modern encounters with unexplained intelligences are different points of contact within the same conscious system. Divinity may be intelligence operating at scales superior to ours, and what we call paranormal is normal from a wider perspective.

Civilizations come and go, but the overall pattern remains

the same because the underlying fabric does. We're part of the universe, witnessing expressions of a greater intelligence trying to understand itself through us.

That may be why every tradition ultimately arrives at the same conclusion: the kingdom of heaven is within; the universe is mental; as above, so below; tat tvam asi—*you are that.*

The search for God, the search for extraterrestrial life, the search for consciousness, and the search for the fundamental laws of physics may all be the same search, seen from different angles. We've been walking around the same mountain, arguing over which side is real.

It's all about Relationships

Consciousness and matter are fundamentally interconnected. The observer and the observed are parts of a unified whole, not separate entities. This is about active participation in a shared reality.

And if that is true, then the veil was never an actual barrier between humanity and the divine. Instead, it was merely the illusion created by our belief in separation, convincing us we were apart from the divine.

The ancients had a better sense of things than we do. They built temples facing the stars because they believed heaven and earth were connected. They considered speech sacred because words could shape the world around them. They were cautious about certain knowledge because they understood awareness held power. They worshipped the gods out of respect for the world they knew they were all a part of.

The universe is not just empty matter in dead space, but something aware, watching, waiting, and perhaps remembering us as we remember it—*the final revelation.*

We are not alone...

Global Organizations Investigating UAP / Non-Terrestrial Phenomena

Around the world, governments, militaries, scientists, and civilian researchers have independently reached the same conclusion: a subset of aerial and transmedium phenomena remains unexplained. Although interpretations vary, ongoing investigations across cultures and political systems indicate that the phenomenon is not confined to a single nation or narrative.

UNITED STATES

Official / Government Programs (Past & Present)

• *Project Sign* (1947–1949). First U.S. Air Force UFO investigation.

• *Project Grudge* (1949–1951). Skeptical continuation of Project Sign.

• *Project Blue Book* (1952–1969). Public-facing Air Force UFO investigation; over 12,000 cases logged.

• *Advanced Aerospace Threat Identification Program (AATIP)*. Pentagon program investigating anomalous aerospace threats.

• *Unidentified Aerial Phenomena Task Force* (*UAPTF*). Naval-led task force focused on military encounters.

• *All-domain Anomaly Resolution Office* (*AARO*). Current DoD office tasked with investigating UAP across air, sea, space, and transmedium domains.

• *US Legacy Programs* (*Unacknowledged / Alleged*). A term used by whistleblowers and researchers to describe earlier classified or compartmentalized efforts predating AATIP. Not officially confirmed, but frequently referenced in testimony.

Intelligence & Research Institutions

• *CIA (historical involvement)*. Collected, analyzed, and classified UFO reports during the Cold War.

• *DIA* (*Defence Intelligence Agency*). Sponsored studies on anomalous aerospace technologies and cognition.

• *DARPA*. Funds frontier research into advanced propulsion, cognition, perception, and sensing (not officially UFO-focused, but adjacent).

• *NASA* (*current UAP research role*). Conducts scientific assessments of UAP data using civilian datasets.

Private & Academic Initiatives (US-based)

• *The Galileo Project* (*Harvard University*) is a Scientific effort to detect and analyze anomalous objects using open data.

• *To The Stars Academy* (*TTSA*) is a Public-private research and media organization focused on UAP disclosure.

• *Scientific Coalition for UAP Studies* (*SCU*) Independent scientists analyzing military and sensor-confirmed UAP cases.

CANADA

• *Sky Canada Project:* Government-supported initiative to modernize Canada's approach to UAP data collection and analysis.

• *National Research Council of Canada (historical role).* Formerly involved in collecting UFO reports on behalf of the government.

• *Transport Canada / DND (indirect involvement).* Aviation safety and defence-related monitoring of anomalous objects.

FRANCE

• *GEIPAN (Groupe d'Études et d'Informations sur les Phénomènes Aérospatiaux Non Identifiés).* One of the world's most transparent government UFO agencies. Operates under the French *space agency* CNES. Publishes case files openly and classifies sightings.

EUROPE (MULTINATIONAL & NATIONAL)

Multinational / Pan-European

• *European Initiative for UAP/NHI Disclosure.* Policy-oriented summit and advocacy initiative pushing coordinated transparency across Europe.

UNITED KINGDOM

• *UK Ministry of Defence – UFO Desk (closed).* Collected sightings until 2009; archives released to the public.

• *Defence Intelligence Staff (historical analysis).* Internal assessments of UFO reports.

SPAIN

• *Spanish Air Force (Ejército del Aire).* Declassified UFO case files related to military airspace.

ITALY

• *Italian Air Force (Aeronautica Militare).* Maintains historical UFO reporting archives.

GERMANY

• *Bundeswehr (limited disclosure).* Military monitoring of anomalous airspace events.

BELGIUM

• *Belgian Air Force (UFO wave investigations)* Notable for transparency during the 1989–1990 wave.

RUSSIA / FORMER USSR

• *Cosmopoisk* is a civilian scientific research organization founded to investigate anomalous phenomena, UFOs, and historical mysteries.

• *Soviet-era Military Research Programs (historical).* Studied UFO encounters reported by pilots and radar operators. Details remain classified mainly.

• *Russian Academy of Sciences (skeptical investigations).* Periodically reviewed anomalous claims.

SOUTH AMERICA

BRAZIL

• *Brazilian Air Force UFO Program.* Official investigations and significant declassification have occurred in recent years.

CHILE

• *CEFAA (Committee for the Study of Anomalous Aerial Phenomena)*

is a Civil-military organization studying UAP with scientific advisors.

PERU

• *Peruvian Air Force – Office of Aerospace Phenomena Investigation.* Monitors reports of anomalous aerial objects.

ASIA

JAPAN

• *The Japanese Ministry of Defence (recent policy shift)* now encourages pilots to report UAP.

CHINA

• *Chinese Military & Space Agencies (indirect).* Publicly acknowledges monitoring of unidentified aerial objects, often framing them as national security concerns.

INDIA

• *Indian Air Force (limited public acknowledgment).* Occasional reports of anomalous aerial sightings.

AUSTRALIA & OCEANIA

• *Royal Australian Air Force (historical UFO files).* Declassified reports are available through the National Archives.

• *Australian Department of Defence (current monitoring).* Passive tracking of anomalous airspace events.

INTERNATIONAL & NON-GOVERNMENTAL

• *National UFO Reporting Center (NUFORC).* Civilian reporting database (US-based).

• *Mutual UFO Network (MUFON)* is an International civilian investigative organization.

• *The International Coalition for Extraterrestrial Research (ICER)* is an Academic-style global research coalition.

• *Scientific Exploration of Anomalous Phenomena* (SEAP) is research into UAP and high-strangeness phenomena.

Epilogue

Harmony flows through all life, matter, and consciousness, echoing the cosmos's symphony, where love blossoms.

The Nested Emergence Hierarchy

The "hidden planetary system" is in the math. What we call Earth is a self-organizing system, shaped by layers of interaction that build on one another. This process is known as *nested emergence.* Larger, more complex patterns arise naturally from smaller ones yet are constrained by the larger system within which they sit.

At the most fundamental level, local interactions take precedence. Air, water, rock, and heat interact to produce weather, tides, and seismic activity. These processes seem chaotic, but they follow specific rules. At a higher level, Earth and the Moon form a coupled system. Gravity keeps them in rhythms that regulate tides, slow Earth's rotation, and stabilize its tilt. Together, they behave like a single resonant structure.

Zooming out further reveals the solar system. The Sun and planets subtly limit which Earth~Moon patterns remain stable and which disappear. Some rhythms persist because the larger system permits them, and others never develop at all.

At the deepest level, there may be universal rules governing space, time, and information that control which large-scale patterns can exist anywhere. Some planetary patterns seem too precise to happen by chance.

Think of it like a conductorless orchestra. Individual instruments play freely, but the room, acoustics, and physical limits of sound shape what the music can become. A coherent symphony erupts.

We could also imagine Russian nesting dolls. The smallest doll represents local Earth systems; the next depicts the Earth~Moon relationship; the third represents the solar system; and the final doll represents the universe. The pattern you observe at the largest scale only exists because each smaller layer fits properly inside the next.

Earth is neither engineered traditionally nor is its existence purely accidental. Instead, it results from multiple layers of interactions that lock into a long-lasting balance. The equations we use are incomplete when explaining how non-local correlations arise across different scales.

This is important because when systems act this way, what we see as control can actually just be a natural outcome. There's no simple switch to turn or machine to find. The system keeps going because physics naturally favours equilibrium, and any instability tends to fade over time.

That's where consciousness functions as another level of manifestation, able to connect with distant structures, just as planetary systems do.

When remote perception works well, it follows the laws of

physics. As systems grow more complex and interconnected, qualities like intelligence, stability, and contact tend to develop naturally.

Bibliography

APPENDIX A

Ancient Civilizations, Megaliths, and Alternative Archaeology

Bauval, Robert & Hancock, Graham. *The Message of the Sphinx.* Crown Publishers, 1996.

Hancock, Graham. *Fingerprints of the Gods.* Crown Publishers, 1995.

Hancock, Graham. *Magicians of the Gods.* Thomas Dunne Books, 2015.

Hancock, Graham. *America Before.* St. Martin's Press, 2019.

Lehner, Mark. *The Complete Pyramids.* Thames & Hudson, 1997.

Dunn, Christopher. *The Giza Power Plant.* Bear & Company, 1998.

West, John Anthony. *Serpent in the Sky.* Quest Books, 1979.

Hancock, Graham & Bauval, Robert. *Talisman: Sacred Cities, Secret Faith.* 2004.

Mythology, Symbolism, and Sacred Traditions

Campbell, Joseph. *The Hero with a Thousand Faces.* Princeton University Press, 1949.

Eliade, Mircea. *The Sacred and the Profane.* Harcourt, 1957.

Eliade, Mircea. *Myths, Dreams and Mysteries.* Harper & Row, 1960.

Jung, Carl. *Man and His Symbols.* Doubleday, 1964.

Jung, Carl. *Archetypes and the Collective Unconscious.* Princeton University Press, 1959.

Armstrong, Karen. *A Short History of Myth.* Canongate, 2005.

Consciousness, Panpsychism, and Philosophy of Mind

Chalmers, David. *The Conscious Mind.* Oxford University Press, 1996.

Goff, Philip. *Galileo's Error: Foundations for a New Science of Consciousness.* Pantheon, 2019.

Sheldrake, Rupert. *The Science Delusion.* Coronet, 2012.

Bohm, David. *Wholeness and the Implicate Order.* Routledge, 1980.

Hameroff, Stuart & Penrose, Roger. Selected papers on Orchestrated Objective Reduction.

Nagel, Thomas. *Mind and Cosmos.* Oxford University Press, 2012.

Quantum Physics and Theoretical Reality

Bohm, David. *Quantum Theory.* Dover Publications, 1989.

Greene, Brian. *The Elegant Universe*. W.W. Norton, 1999.
Greene, Brian. *The Fabric of the Cosmos*. Knopf, 2004.
Barrow, John D. *The Constants of Nature*. Vintage, 2003.
Kaku, Michio. *Parallel Worlds*. Doubleday, 2004.
Rovelli, Carlo. *Reality Is Not What It Seems*. Riverhead, 2017.
Wheeler, John Archibald. Selected essays on observer-participancy.

Earth Resonance and Planetary Systems

Schumann, Winfried Otto. Foundational papers on electromagnetic resonance (1952–1957).
Helliwell, Robert. *Whistlers and Related Ionospheric Phenomena*. Stanford University Press, 1965.
Lakhovsky, Georges. *The Secret of Life*. 1929.
Persinger, Michael A. Selected research on geomagnetic activity and consciousness.
Voll, Reinhold. Electroacupuncture and bioelectromagnetics research.

Intelligence, Propaganda, and Psychological Operations

Bernays, Edward. *Propaganda*. 1928.
Ellul, Jacques. *Propaganda: The Formation of Men's Attitudes*. Vintage, 1965.
Herman, Edward & Chomsky, Noam. *Manufacturing Consent*. Pantheon, 1988.
Sun Tzu. *The Art of War*. Various translations.
McLuhan, Marshall. *Understanding Media*. MIT Press, 1964.
Marks, John. *The Search for the "Manchurian Candidate"*. Times Books, 1979.
U.S. Central Intelligence Agency. MKULTRA Declassified Documents. National Archives.

UFO Phenomena and Anomalous Contact

Vallee, Jacques. *Passport to Magonia*. 1969.
Vallee, Jacques. *Dimensions*. 1988.
Hynek, J. Allen. *The UFO Experience*. 1972.
Keel, John. *The Mothman Prophecies*. 1975.
Mack, John E. *Abduction*. Scribner, 1994.
Strieber, Whitley. *Communion*. 1987.
Friedman, Stanton T. Lectures and research papers on UFO investigations.

Near-Death Experience and Mysticism

Moody, Raymond. *Life After Life*. 1975.
Ring, Kenneth. *Life at Death*. 1980.
Grof, Stanislav. *The Holotropic Mind*. 1992.
Tolle, Eckhart. *The Power of Now*. 1997.

Bibliography

Watts, Alan. *The Book on the Taboo Against Knowing Who You Are*. 1966.
James, William. *The Varieties of Religious Experience*. 1902.

Sacred Texts and Religious Foundations
The *Rig Veda*. Various translations.
The *Upanishads*. Various translations.
The *Bhagavad Gita*.
The *Bible* (King James Version and modern scholarly translations).
The *Nag Hammadi Library* (Gnostic Gospels).
The *Zohar* (Kabbalistic texts).
Laozi. *Tao Te Ching*. Various translations.

Interdisciplinary and Synthesis Works
Capra, Fritjof. *The Tao of Physics*. 1975.
Prigogine, Ilya. *Order Out of Chaos*. 1984.
Talbot, Michael. *The Holographic Universe*. 1991.
Wilber, Ken. *A Brief History of Everything*. 1996.
McKenna, Terence. *Food of the Gods*. 1992.
Sheldrake, Rupert. *Morphic Resonance*.

About the Author

Jody Swannell, an author from Ontario, Canada, is passionately interested in exploration, both physical and intellectual. Her work is driven by a lifelong curiosity about the world and ideas that challenge traditional views.

Her travels have taken her from Scotland's historic castles to Alaska's glaciers, the Caribbean, the United States, and across Canada. These experiences have shaped her perspective, grounding her writing in both observation and inquiry.

Jody writes in various genres, including thrillers, science fiction, and stories about the occult and mysterious phenomena. Her nonfiction work especially shows her dedication to exploring complex and controversial topics with an open, curious mindset, inviting readers to think carefully about what's real and how much we truly know.

instagram.com/swannellbookery

facebook.com/JodySwannell42

tiktok.com/@jodyswannell

Made in the USA
Monee, IL
04 November 2026

67383133R00154